Praise for
FIRE AND BRIMSTONE

P9-DXK-720

"An absorbing story...Punke, clearly at ease writing about politics, Washington, and the law, uses the strike and the disaster's aftermath to reveal the circular nature of corruption in company towns. It is difficult to read these passages and not think of the Sago mine disaster in West Virginia and not wonder why in the near century since, safety for miners had still not improved significantly in the US."

—*Los Angeles Times Book Review*

"Punke presents this timely story in a strong and readable manner."

—*Library Journal*

"Reads with the immediacy of today's tragic headlines. *Fire and Brimstone* combines a driving narrative energy with the authority of detail only the most careful research can reveal."

—William "Gatz" Hjortsberg, author of
Odd Corners and screenwriter of *Legend*

"Punke's recounting of the struggle of the others to survive is tense, exciting, and even inspiring. He tells an equally interesting story of the ethnic conflicts, anti-war protests, and labor warfare."

—*Booklist*

"Like the North Butte fire itself, Michael Punke's account races down a dark tunnel in our nation's history. It is a gripping read, both as a heart-pounding story of tragedy and heroism in the mines, and a deeper reflection—so relevant today—on how fear-mongering, parading under the name of patriotism, can pervert our nation's most cherished values."

—Peter Stark, correspondent, *Outside*
magazine and bestselling author of *Astoria*

CALGARY PUBLIC LIBRARY

SEP   2016

FIRE AND BRIMSTONE

FIRE AND BRIMSTONE

•

*The
North Butte
Mining Disaster
of 1917*

•

MICHAEL PUNKE

NEW YORK   BOSTON

Copyright © 2006 Michael Punke

Source for diagram on page xii:
Bureau of Mines, "Lessons from the Granite Mountain
Shaft Fire, Butte," 1922.

All rights reserved. In accordance with the U.S. Copyright Act of 1976, the scanning,
uploading, and electronic sharing of any part of this book without the permission of
the publisher constitute unlawful piracy and theft of the author's intellectual
property. If you would like to use material from the book (other than for review
purposes), prior written permission must be obtained by contacting the publisher at
permissions@hbgusa.com. Thank you for your support of the author's rights.

Hachette Books
Hachette Book Group
1290 Avenue of the Americas
New York, NY 10104
www.HachetteBookGroup.com

Printed in the United States of America

RRD-C

Originally published by Hyperion.
First Hachette Books trade paperback edition: February 2016

10 9 8 7 6 5 4

Hachette Books is a division of Hachette Book Group, Inc.
The Hachette Books name and logo are trademarks of Hachette Book Group, Inc.

The publisher is not responsible for websites (or their content) that are not owned
by the publisher.

Library of Congress Cataloging-in-Publication Data

Punke, Michael.
Fire and brimstone : the North Butte mining disaster
of 1917 / Michael Punke.— 1st ed.
    p.   cm.
Includes index.
ISBN 978-1-4013-0889-6
1. Mine fires—Montana—Butte.   2. Granite Mountain
Speculator Mine Disaster, Mont., 1917.   3. Copper
mines and mining—Accidents—Montana—Butte.
4. North Butte Mining Company.   I. Title.
TN315.P86 2006
363.11'9622343'0978668—dc22
2005052579

Dedicated to the people of Butte, Montana

CONTENTS

Contents

Tell her we done the best we could,
but the cards were against us.
—J.D. MOORE,
JUNE 9, 1917

FIRE AND BRIMSTONE

CROSS SECTION OF SPECULATOR
AND GRANITE MOUNTAIN SHAFTS

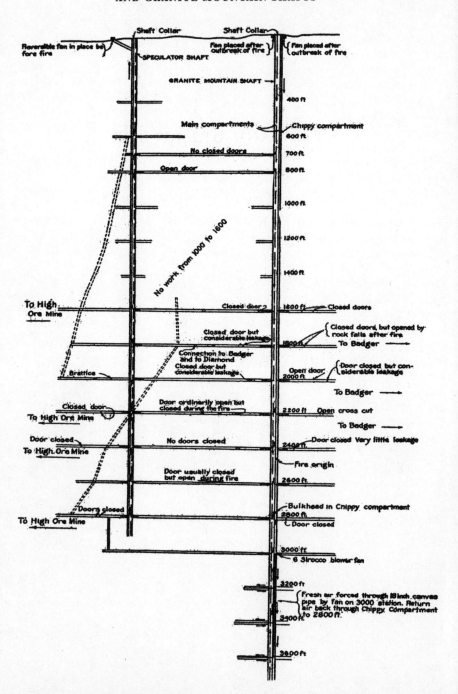

One

•

"THERE IS A SIGN"

There is a sign that appears to point persistently
to a terrible explosion underground.
—HOROSCOPE PRINTED IN THE
ANACONDA STANDARD, JUNE 5, 1917

Butte, Montana, was still dark when the cable crew arrived at the Granite Mountain shaft in the early morning hours of Friday, June 8, 1917. The men could see their breath as they worked in the cold, open air beneath the headframe—a hulking, eight-story steel structure that supported the massive weight of the hoists. The miners called it a "gallows frame" because it looked like a gigantic industrial version of the Old West hangman's platform. Gallows frames dotted the sprawling hillside like Butte's version of trees.

The official accident report does not list the names of the men on the crew that morning, but it is a good bet they were a motley mix—the NO SMOKING signs that hung around the mine shaft were printed in sixteen languages. The report does provide the men's job titles, titles that reflect the task before them: "four electricians, three rope-men, two shaft-men, and one hoist-man."[1]

The worst hard-rock mining disaster in American history began, ironically, as an effort to improve the safety of the Granite Mountain shaft. Granite Mountain, along with its sister shaft, the Speculator, were owned by the North Butte Mining Company. The two mines were rarities—productive Butte properties *not* owned by the omnipotent Anaconda Copper Mining Company. Anaconda did own all of the properties that surrounded the North Butte holdings. Anaconda, indeed, owned most of the city and a sizable chunk of the state.

By the lax standards of the day, the North Butte Mining Company boasted a considerable reputation for safety. The cable crew's work on the morning of June 8 was part of an effort to install a sprinkler system up and down Granite Mountain's 3,700-foot shaft. The shaft was buttressed with chemically treated wooden timbers. In a fire, this highly flammable lining was the functional equivalent of a chimney made of wood. The new sprinkler system, though, was nearly complete. In a week or so, a shaft fire could be doused by simply turning a valve.

One of the few tasks remaining was to relocate a large electrical transformer at the 2,600-foot Station. "Stations" were cavernous openings at hundred-foot intervals and were the junctions between the main shaft and the hundreds of tunnels that branched out from it. The 2,600 Station, therefore, was 2,600 feet below the surface, or "collar," of the shaft.

The transformer at the 2,600 Station stood only fifty feet from the wood-timbered shaft. Worried that an electrical fire could easily spread, mine officials had decided to move the transformer to a safer location—deeper inside the workings of the mine and well away from the main shaft. The electrical cable that connected it to the surface, however, was too short. The job assigned to the crew was to lower a long cable that could reach the transformer's new location.

In the era before plastic, electrical wire was commonly insulated with oil-soaked cloth, usually cambric or jute. In industrial settings such as a mine, the cloth was then sheathed in lead for protection. The new 1,200-foot electrical cable being lowered into Granite Mountain was a full five inches in diameter and weighed five pounds *per foot,* putting its total mass at a staggering three tons.

The narrow five-by-sixteen-foot main shaft contained three hoists—side-by-side elevator cages that could be operated independently of one another. The cages were controlled by a "hoist-man" on the surface. Two of the hoists were dedicated to pulling up lodes of copper ore from the depths of the mine. The other cage, called a "chippy," was used to transport men and materials. The men in the shaft's various stations communicated with the hoist-man through a complicated system of electric-powered bell signals.

To lower the new cable into the Granite Mountain shaft, the crew planned to use the mechanical strength of the third hoist, the chippy. This necessitated the labor-intensive act of lashing the cable to the chippy's lowering rope (the rope on which the cage itself was pulled up and down). As the first 500 feet of cable was lowered, the crew tied the cable and the lowering rope together with hemp lashes at ten-foot intervals. After the first 500 feet, the crew tied lashes every *five* feet to compensate for the ever-increasing weight.[2] By the time the entire cable was tied to the chippy's rope, nearly 200 lashes were in place.

The crew had worked for sixteen straight hours before the lower end of the cable finally came even with the 2,600 Station. Sometime before 8:00 P.M., the men were about ready to begin threading the cable toward the site for the new transformer when they noticed a problem. The bottom 200 feet of cable appeared to be "coiled around the hoisting rope." Before snaking the cable into the 2,600 Station, the crew decided to make an attempt at straightening out the kinks.

One of the men came up with the idea of removing some of the lower lashings. The hope was that the three-ton cable, if freed from its lower bindings to the hoisting rope, would untwist itself—the way kinks unwind from a tangled telephone cord.

The details of how the crew went about releasing the hemp ties are vague. It appears, however, that two or more of the men scaled the shaft walls and began cutting or untying the lashings. Adding to the precariousness of their task was the fact that the hoisting rope was smeared with lubricant to help it spool and unspool more easily.

Whether or not the kinks began to unwind is unclear. What *is* clear is that as the men of the cable crew worked around 8:00 P.M. on Friday night, they noticed that the three-ton cable suspended above them was slipping. Then, in a heart-sinking moment, they realized it was falling down.

The crew made a mad scramble for cover and "had scarcely reached the safety on the station before the cable fell." As they huddled in the 2,600 Station—which opened directly to the shaft—the men heard and felt the thunderous roar of the falling cable. It bounced and scraped for hundreds of feet along the timbers of the narrow shaft, ripping every impediment in its path. Air hoses and water pipes were pulled off their brackets, adding the high-pitched scream of twisting metal. Finally the cable became caught in the shaft, piling up in a giant, tangled clump between the 2,400 and 2,800 levels—above, below, and beside the cringing cable crew at the 2,600 Station.

There must have been a palpable pause when the crashing finally stopped, one of those suddenly quiet, postcalamity moments in which accident victims take shocked inventory of themselves and their surroundings. Eventually the crew moved hesitantly toward the shaft, their lanterns barely able to penetrate the thick dust thrown up by the falling cable.

The cable was ruined. The violent fall had scraped away half of the lead sheathing, exposing large sections of the oil-soaked cloth insulation. After they inspected the snarled pile, it was obvious to the men that the giant wire could no longer carry an electrical current.

By then it was 10:00 P.M., and the crew had been working for eighteen straight hours. In no mood to begin the difficult task of removing the ruined cable, they headed for the surface.

B urton K. "B.K." Wheeler, Butte's thirty-five-year-old federal district attorney, was a man in an unenviable position. Wheeler's job was to enforce federal law, and Butte, Montana, on the eve of the North Butte disaster was a volatile, messy jumble of antiwar protest, an abusive corporate master, seething labor unrest, divisive ethnic tension, and radicalism both left and right. It was a powder keg lacking only a spark, and in the days surrounding the disaster, multiple conflicts would converge and explode. B.K. Wheeler would ride through the center of this firestorm, then carry its legacy in the decades to follow.

Only two months earlier, in April 1917, the United States had waded into the bloody fray of World War I. For Butte, one immediate effect of the conflict was the widening of ethnic fissures. Butte was a microcosm of Europe, and Europe was at war. German immigrants opposed fighting against their recent homeland, where many of their families still lived. The English and the French, by contrast, cheered America's alignment with their countrymen. The Irish, with historical animosity toward England, stood in bitter opposition to an American alliance with the British. The Finns, strongly socialist, saw the war as a scheme to "break the power of the people of Russia."[3] War also led to an increase in immigration from eastern and southern Europe, and all of Butte's then-current inhabitants resented the influx of Italians and Slavs, who as the newest wave of immigrants were willing to work for the lowest wages.

Leftists marched in opposition to the "rich man's war, poor man's fight" and plotted the downfall of the capitalist system. Rightists launched a hunt for "Shadow Huns" and changed the names of German foods. In the approved parlance, *hamburger* became *Salisbury steak, frankfurters* became *hot dogs,* and *sauerkraut* became *liberty cabbage.*[4]

An early flashpoint in Butte centered on registration for a new draft, with Registration Day falling on June 5, 1917—only three days before the North Butte disaster. Butte war opponents, led by the Finns and Irish, circulated handbills screaming "WAR IS HELL. WE DO NOT WANT IT" and "DO NOT REGISTER." "[W]e are," warned the handbills, "at the behest of the money powers, to be taken forcibly to kill and be killed."[5] The draft even threatened to reignite problems with the Indians. Butte newspapers ran a wire story reporting that antiwar protesters included the Cheyenne, who were "holding war dances and threatening violences."[6]

The responsibility to enforce the draft law seemed an odd match for District Attorney B.K. Wheeler. As a Quaker, he had his own considerable misgivings about American involvement in the war and had established his young career by representing Butte's working-class men—usually against "the money powers." Yet as the federal district attorney, Wheeler took seriously his responsibility to enforce the law and issued a tough statement on the eve of Registration Day. "Any man within the draft age who is heard making the remark that he will not register will be warned during the day by the Attorney's force. If he has not registered by nine p.m., he will be taken promptly to jail."[7]

Wheeler's tough tone dissuaded some but not all. On Registration Day, a group of Finns and Irish led a protest march of as many as 2,500 people. Antiwar speeches were delivered in English and in Finnish before the Butte mayor addressed the crowd (from the top of a building), demanding it disperse. The order was met with jeering and boos that quickly degenerated into rioting, leading police and sheriff's deputies to fire shots into the air and then wade into the crowd with clubs and long nightsticks. When the rioters still failed to disperse, the National Guard was called in. Forty soldiers with rifles and fixed bayonets "came at a run" from an encampment on the edge of town. Finally the riot broke up, with twenty men and one woman arrested.[8]

"The draft riots were forgotten three days later when fire broke out," remembered Wheeler.[9] Forgotten for a time, though the riot offered a preview of the broader ramifications of the disaster that was about to erupt—a disaster that would envelop Wheeler's long career in ways that he could not possibly foresee.

A s they ascended the shaft, the members of the cable crew no doubt felt a mixture of relief and trepidation. They knew they were lucky to be alive, but there would be serious consequences for their role in ruining a valuable piece of mine property. The cable cost more than $5,000—the equivalent of a year's wage for four men.[10]

What the cable crew could not know was that the events they had set in motion would stretch far beyond a $5,000 electrical wire. In less than two hours, hundreds of men would be locked in a desperate struggle to survive. In less than three days, 163 of them would be dead.[11]

Nor would the death and destruction be limited to the mines.

Before the last chapter was written, the legacy of the disaster would include murder, a crippling strike, an ethnic and political witch hunt, a national law effectively suspending the First Amendment, and an epic battle over presidential power. Butte, Montana, sits in the heart of the American West—but this is the story of a very different frontier.

Daylight had come and gone when the cable crew reached the surface, late in the evening of June 8, 1917. The men reported the lost cable to Ernest Sullau, the assistant foreman. Sullau had just arrived at work, and the responsibility for pulling the ruined cable from the shaft would fall to him. Mine operations ran round the clock, and Sullau headed up the graveyard shift.

Two

•

"LIKE A GIGANTIC TORCH"

An appalling sight which caused the strongest hearts to quail was the cremation of two men, Mike Conway [sic] and [Peter]Sheridan, station tenders, who were trapped like rats in a double decked cage, about twenty feet above the collar of the shaft, with the flames flying from the shaft like a gigantic torch around them.
—*BUTTE MINER*, JUNE 9, 1917

The miners who worked for forty-eight-year-old Ernest Sullau called him "Sully." Just about everybody in Butte, it seems, had a nickname.

Sullau was born in Hamburg, Germany, on Christmas Day 1868. He came to America as a "small boy" and spent most of his adult life as a miner, including stints chasing gold in Klondike and Nome. When he arrived in Montana around 1897, he first worked as a placer miner—sifting small claims of his own. By 1900, though, Sullau gave up his quest for the big strike, opting instead for the steady wages of industrial mining. By 1917, he was a seasoned, fifteen-year veteran of the North Butte Mining Company's Speculator and Granite Mountain mines.[1]

While Sullau never got rich, the stability he found in Butte's mines brought him other benefits. In 1911, at the age of forty-two, Sullau married Lena Benson, a woman unabashed in her affection for her husband. "He was the best man that ever lived," she said. "When the twenty-to-six car came I knew he would be on it." Through his hard work and experience with the North Butte Mining Company, Sullau had risen to the rank of assistant foreman. Between his respected position in the mine and his comfortable home life, Sullau had reached a "good place in life."[2]

Sullau's task on the evening of June 8, 1917, was straightforward: Descend the mine in the auxiliary cage (which was not blocked by the ruined cable); find the top end of the cable; attach it to a hoisting rope; and pull it up. With the cable removed, an assessment could be made of the damage done to the shaft.

Already it was clear that the water supply to the lower reaches of Granite Mountain had been severed. Without water, the miners' Leyner drills could not function. The men at the 3,000 level, unable to work, had gone home, no doubt grumbling about the wages they lost in a workday cut short.[3]

Accompanying Sullau into the mine were a shift boss named John "Baldy" Collins and two shaft men. It was 11:30 P.M. when the four men crowded into the open cage and began their descent. Their destination was a point fifty feet below the 2,400 Station—where the top portion of the electrical cable had lodged.

The ride down, covering nearly half a mile, would have taken around five minutes. When the men finally came even with the tangled mass, there was no sign of the end of the cable. To search for it, Sullau and Baldy crawled *out of the cage*—edging perilously along the timbers on the side and center of the shaft. Sullau went one direction and Baldy went the other—with the web of cable in between them.[4]

The Granite Mountain shaft measured 3,740 feet from top to bottom—the deepest in Butte. Though already 2,450 feet below the surface, the men were still a gut-churning 1,300-foot drop above the bottom of the shaft. Their position was the equivalent of hanging from the top of the Empire State Building—in the dark.

As Sullau gripped the shaft timbers, he also held tight to the carbide-burning lantern that every miner was required to buy. Though electric headlamps had been available since the turn of the century (and were widely used in coal mines), carbide lanterns saw far wider use through the 1920s. Improvements to electric lamps in the 1930s would make them brighter and less heavy, but before then, the weight of the batteries made them so bulky that miners didn't like to carry them.[5]

So as Ernest Sullau searched for the end of the ruined cable at 11:45 P.M., he held in his hand an open flame—not even a pane of glass enclosed the fire. The environment could hardly have been more volatile. The fall of the giant cable had scraped away half of its lead sheathing, exposing large sections of the highly flammable, oil-soaked cloth that insulated the wire. Boys who lived near the mines picked up scraps of this same type of electrical cable, stripping them to sell for the copper. The boys used their pocketknives to peel off the lead. To clean away the cloth, though, they simply touched it with a match. It burned "like gasoline."[6]

Somehow, as Sullau clung to the timbers, probing into the darkness, his lantern came into contact with the exposed cloth on the cable. It burst into flames. Reflexively, Sullau recoiled his lantern arm—sparking a second set of flames behind him.

"John!" Sullau yelled to Baldy. "I got my lamp too close!"

Baldy clambered around to Sullau's position and the two men worked furiously to stamp out the first set of flames. For an instant it seemed to work. But then, remembered Baldy, the fire "immediately broke out in the cable on the opposite side." Furiously they flailed at this second outbreak, but the fire by now was taking on a life of its own.[7]

The flames spread rapidly through the tangled cable, the frayed wire like a 1,200-foot fuse. To make matters worse, the Granite Mountain shaft had a powerful downdraft. Under normal circumstances, air circulated down the Granite Mountain shaft, then traveled sideways through numerous "crosscuts" to its sister shaft, the Speculator. The Spec, an updraft, drew the air back up to the surface. Air in this system moved at a breezy 200 to 300 feet per minute, making the Granite Mountain and Speculator shafts among the best ventilated in Butte.[8] This circulation, normally beneficial, now turned deadly. In the initial moments after flames broke out, the Granite Mountain downdraft sucked the fire into the mass of cable, then fanned the growing conflagration.[9]

Sullau and Baldy, fearful for their lives, scrambled back in the cage, signaling the surface furiously to be lifted up to the 2,400 Station. The cage responded at once—pulling them up fifty feet to the station. A blunt-faced, big-handed miner named Albert Cobb was one of several men working at the 2,400 that night. "I had just pulled out of the 2,400 of the Granite Mountain with a load of rock."[10]

Sullau threw open the cage door and jumped out. Baldy, remaining in the cage, signaled the surface to pull him and the two other men to the top. The cage disappeared as Sullau yelled desperately to Albert Cobb, "Give me your bucket quick."[11]

Another man in the station knocked the top off of a ten-gallon drinking keg and carried it toward the shaft. The fire, though, was already far beyond buckets. "Before we could get over to the shaft," Cobb would later report, "the smoke was coming up in frightful fashion."

The heat of the flames and the smoke they threw off reversed the Granite Mountain draft. The downdraft, which had initially sucked the fire downward, turned updraft. Now pulled from above, the fire began to race up the half mile of shaft above the 2,400 Station—feeding on the tinder-dry, chemically treated timbers.

Sullau began to confront the enormity of his mistake. "For God's sake, get the men out!" he yelled. "Get 'em out!"[12]

B aldy Collins and the two other men reached the surface and piled out of the cage. The fire, they knew, would burn quickly up the wooden shaft. Speed was essential if they were to save the men working below. Desperately they set to work, attaching additional cages to the hoist lines so that more men could be lifted at a time.[13]

Two of the men assisting Baldy on the surface were Peter Sheridan and Mike Conroy. Sheridan and Conroy were "station tenders" responsible for the 2,200 Station. A fire, they knew from Baldy, had broken out only a few hundred feet below the 2,200—*their* station, where *their* men were at work. The tenders argued with Baldy to be lowered down in an effort to bring the men up. Baldy refused, worried now that the hoist's electric signaling system might not be working. He ordered the two men to stay on the surface until he had a chance to consult with the engineer who operated the hoist—two hundred feet away, out of sight, on the inside of the engine room. Baldy turned and ran for the engine room.

Before Baldy could reach the engineer, Sheridan and Conroy took matters into their own hands, signaling to be lowered to the 2,200 Station. A third man, Con McLafferty, jumped onboard with them.[14] The cage dropped into the shaft.

The Granite Mountain shaft had four electric signal systems through which the men on the surface could communicate with the men in the depths of the mine. By midnight—only fifteen minutes after Ernest Sullau's lantern set the cable ablaze—the roaring shaft fire had destroyed all four.[15] The communication system was critical to the operation of the hoists, and the severance of signal lines had immediate and horrific consequences.

By the time Sheridan, Conroy, and McLafferty reached the 2,200 Station, the fire was close enough to the cage that McLafferty jumped out and "refused to take the chance of going up." Instead, he retreated into the mine workings at the 2,200 level, ultimately surviving to tell the tale.[16]

Sheridan and Conroy must have calculated that they could still outrun the flames aboard the hoist. Like hundreds of other men at work in the mine that night, the two station tenders made snap judgments that sealed their fate. To signal the engineer on the surface to pull them up, they would have attempted to ring the bell—unaware that the ruined electrical wires kept their desperate signal from reaching the surface. By now the fire was virtually exploding up the shaft. Within a few instants, Sheridan and Conroy were enveloped in flames.

On the surface, the engineer, with Baldy at his side, became increasingly uneasy.[17] From the station where he operated the levers that raised and lowered the hoists, the engineer watched a giant gauge that showed the position of the cage—sitting at the 2,200 Station. But there had been no signal to raise it. The miners communicated with the surface in a sort of abbreviated version of Morse code. In each station was a cord with a large handle. Pulling the cord completed an electrical circuit, ringing a bell in the hoist house. There was a specific call to indicate that the cage was clear: *two bells, pause, one bell, pause, two bells*. Once the engineer heard this signal, he knew it was safe to move the cage. In those minutes around midnight, though, he heard nothing from the men on the 2,200.

At about that time smoke started to pour from the Granite Mountain shaft. Frantically, the engineer raised the cage.[18]

By the time it emerged from the collar, the cage was surrounded by fire "flying from the shaft like a gigantic torch." The engineer stopped the cage about forty feet above the surface, dangling from the gallows headframe atop a "geyser of flames." Horrified witnesses could see the charred remains of Sheridan and Conroy, locked in a "death embrace."[19]

In its coverage of the incident, the *Butte Miner* attempted to reassure its readers with the assertion that "the terrific heat without doubt quickly snuffed out the lives of the two station tenders," though it also noted that the victims were found "with arms and feet burned away." Other accounts were even more graphic. The *Butte Daily Post* reported that Sheridan and Conroy were "burned almost to a crisp" and that "even the hair and flesh on their heads [was] gone, so that their skulls were laid bare."[20]

The flames and heat from the fire made it impossible for onlookers to approach the cage. Eventually enough water was poured into the shaft so that two firemen, Angus McLeod and George Lapp, could at least remove the bodies.

With the fire and death at the mouth of Granite Mountain, few onlookers would have noticed what happened at 12:10 A.M., when smoke began to rise from the *Speculator* shaft—800 feet away.[21]

E rnest Sullau could not see the Speculator shaft from his position near the Granite Mountain's 2,400 Station. He didn't need to. As a veteran of the North Butte Mining Company, Sullau understood intuitively the interconnection between the Granite Mountain and the Spec.

Butte sat atop a labyrinth made up of *thousands of miles* of underground mining works.[22] The North Butte's holdings illustrate a small portion of this subterranean world. Its two main shafts—Granite Mountain and the Speculator—ran straight down, each with its own headframe and hoist system to transport men and ore. The two shafts stood 800 feet apart, connected by at least ten crosscut tunnels. If the mine is compared to a human body, the shafts and crosscuts were like main arteries. Branching off of these arteries were hundreds of smaller veins and capillaries. There were "drifts"—horizontal excavations to remove ore; "stopes"—excavations in the shape of a step; and "manways"—man-size, vertical shafts with a ladder inside that allowed men to climb up and down between different levels.[23]

And all of that was just the property of the North Butte Mining Company. The Granite Mountain and the Spec connected directly, at multiple levels, to five other mines, which were owned by Anaconda.

Ernest Sullau knew that the deadly consequences of the fire—his fire—would not be contained in one shaft, or even in one mine. The smoke and fumes spewing forth from the Granite Mountain shaft would spread throughout the far-flung recesses of this underground world, where hundreds of men were at work. Sullau also knew something else—that the fumes were filled with deadly gas.

There is no doubt that at this juncture, minutes after the start of the fire, Ernest Sullau could have saved himself. Though the main shaft was ablaze and its hoist no longer a viable option, other clear avenues of escape were within easy reach.[24] With his long years of experience, Sullau knew the mine workings as well as the streets surrounding his own home.

There is no record of his thoughts as Ernest Sullau made the decision that sealed his own fate that night, but we know he did not run away.

Three

•

"THE RICHEST HILL
ON EARTH"

No one knows it yet, but I have discovered the richest hill on earth.
—ATTRIBUTED TO COPPER KING MARCUS DALY

In the spring of 1856, a small party led by Caleb E. Irvine camped on a high plain below a distinctive butte near the headwaters of the Clark Fork River. They were traders, not miners, but they made a discovery that hinted at the future of the hill on which they stood. Near their camp, Irvine discovered a deep trench, dug into an outcropping of exposed quartz. Next to the trench were elk antlers, apparently used as gads to scrape out the hole.

Who dug the hole and what if anything they found were never discovered. As Irvine and his men pondered the mystery, a band of hostile Blackfoot Indians approached their position, and the traders beat a quick retreat toward the more hospitable country of the Shoshoni and the Crow.[1]

The real discovery of Butte would come as a consequence of the decline in California's supply of "placer" gold. Placer gold was "poor man's diggings"—dust and nuggets close to the surface, ready to be scooped up by methods of no greater sophistication than a metal pan or a wooden sluice. As California's gold industry shifted from placers to more industrialized forms of removal in the mid-1850s, the "forty-niners" began to search for new prospects. They turned inland and by the early 1860s had developed placer strikes in Idaho and Colorado.[2]

In 1862, as the Civil War raged back east, wandering argonauts turned up the first substantial gold strike in what is now Montana. The location of the gold was a small rivulet called Grasshopper Creek. Within a year, miners unearthed an even larger deposit about seventy miles away at a site they dubbed "Alder Gulch." Now the real rush was on. Eighteen months later, ten thousand men were sifting every creek bed between the boomtowns of Bannock and Virginia City.[3]

Among the young miners who joined the gold rush to Montana was a skinny, conniving, blindly ambitious twenty-four-year-old named William A. Clark. As much as any man, Clark would rise to shape a critical, rollicking epoch of western history. His origins, though, were humble; he was born of Scotch-Irish parents on a Pennsylvania farm. The Clark family moved to Iowa, where William continued a solid education that included two years of legal study at Iowa Wesleyan College.[4]

Clark worked as a schoolteacher in Missouri and apparently fought briefly with the Confederate Army before deciding to seek his fortune in the West. Prospects already were fading by the time the young man reached the Colorado mines in 1862, and after working as a "$2.50 a day laborer," Clark headed for the new strikes in Montana. With a partner, he staked a claim near Bannock. They built a log cabin and sluices and, after only a year, sold their property for a healthy profit of $2,000 each. For Clark, this initial stake would grow into a fortune of $47 million and make him one of the richest men in the world.

Clark's early wealth, though, would not come from gold—at least not directly. Like many financial giants of the mining era, Clark's initial success lay not in the mines—but in the miners. With his $2,000 in profits from sluicing gold, he bought mules and began moving freight between Salt Lake and the Montana boomtowns, carrying shovels, picks, tobacco, and even eggs. His goods, according to Clark, sold for "extraordinarily high prices."[5] With his profits he moved into wholesaling, mail transport, and then banking. By the age of thirty-three he was a millionaire and one of the leading financiers in what had become the Territory of Montana. "[E]verything he touched seemed to turn to gold."[6]

"Turn to copper" might have more accurately described Clark's career, for it was not the yellow metal, but the red, that would ultimately create the great bulk of his fortune. Nor was it the soon-to-be ghost town of Bannock with which Clark's name would forever be tied—but Butte.

The first true mining in what was soon to become Butte occurred in the summer of 1864, as Bannock and Virginia City miners began to spread out into untapped areas. Two men, named William Allison and G. O. Humphreys, decided to dig deeper in the shallow pit—the one beside the elk-horn gads—discovered by trader Caleb Irvine eight years earlier. Allison and Humphreys carried dirt from the pit down to Silver Bow Creek to wash it, and in their pans they found gold. They named their site "Baboon Gulch," though this soon gave way to a more stately appellation reflecting the distinctive nearby hill. "Butte City" was born. ("City" was later dropped.)[7]

By 1867, five thousand miners were spread out below the butte and along the meandering Silver Bow Creek. Butte's naissance as a gold mining camp alloyed the town with characteristics from which it would never separate—not that it wanted to. It was a harsh and dangerous place, where efforts to earn a living came often at the cost of men's lives. An eastern reporter, as horrified as his modern counterparts by the West's attachment to firearms, had this to say about the camp: "Every businessman in Butte and every miner is a walking arsenal. He carries a brace or two of pistols in his belt and a Bowie knife in his right boot."[8]

The harsh winters and unsanitary living conditions created a breeding ground for disease, and in mining camps such as Butte "it was a fair estimate that ten percent would die between the months of September and May." Frozen ground made it difficult to bury the dead, so log stockades were sometimes constructed to protect the bodies from the indignity of consumption by wild animals.[9]

With the possibility of death never far removed, miners seized zealously upon the crude opportunities for frontier recreation—centered around the unholy trinity of alcohol, gambling, and prostitutes. Gold might remain elusive for some miners, but the staples of vice seemed to spring up from the soil like weeds. In the vast literature about Butte, no description is repeated more often than "wide open." The same eastern reporter who blanched at Butte's walking arsenals had this description of early Butte entertainment: "Bronchos are ridden into public and private houses as it suits their drunken riders. Men, women, negroes, Chinese and Indians daily and nightly congregate in one common assemblage around the gaming tables with which the dissolute, hilarious camp abounds."[10]

If there is exaggeration in the reporter's description, it most likely concerns the degree of tolerance it implies. Butte would become a melting pot of remarkable ethnic diversity, but "negroes, Chinese and Indians"—while certainly present—were not among those welcomed into the bubbling stew. Butte's first hanging, for example, took place when a miner named Dan Haffie decided to lynch a Chinese "just for luck."[11]

As in most placer camps, the presence of Butte's easy gold did not last long. By 1870, Butte's population had dwindled to 241 souls—98 of whom were Chinese. The high percentage of Chinese indicated the perceived poor quality of the workings. Throughout the West, Chinese often found opportunities in the abandoned claims of less patient miners.[12]

Among those few who stayed in Butte, interest by the late 1860s had begun to shift from gold to another precious metal—silver. The presence of silver had long been obvious in Butte. Black "reefs" pushed through the surface in many places, a clear indication of the potential wealth below. Silver, though, required a different type of mining and a different type of miner. Unlike placer gold—which could be scooped up in pure form by a man with a pan—silver required considerable industry. Deep shafts had to be sunk in the rocky ground. Milling was necessary to crush the ore into a more workable form. And it took smelting to separate the silver from crushed ore. Butte's silver, though plentiful, was notoriously "rebellious," meaning the silver itself was difficult to extract from the surrounding rock. In short, the production of silver required technology and capital.[13]

It was during these early days of Butte's silver mining that the merchant-banker William A. Clark stepped back into the picture. By the early 1870s, the base of Clark's burgeoning frontier enterprises was a bank in Deer Lodge, Montana, less than forty miles from Butte. When he visited the remnants of Butte in 1872, his remarkable eye for investment told him that there was great potential in the largely abandoned town. Clark bought four claims outright and would later begin financing the operations of other silver enthusiasts. Demonstrating an attribute that served him well throughout his life, Clark embraced change. With samples from his new Butte properties, he traveled to New York City and enrolled at the renowned Columbia University

School of Mines. At age thirty-three, William Clark was about to launch a new career in silver mining.[14]

Silver would bring Butte back to life. And with silver would come critical components of the later copper industry: capital, technology, and a stout Irishman who would soon lock horns with William Clark in one of the most dramatic feuds in American history.

Marcus Daly was born in 1841 in Ballyjamesduff, County Cavan, Ireland. Like his enemy William Clark—in fact, to an even greater degree—Daly's life story reads like a Horatio Alger tale. As a child, his family endured the hardships of the potato famine, and young Marcus watched his countrymen abandon the Emerald Isle by the drove.[15]

At the age of fifteen, Marcus emigrated—alone and penniless—to New York City. In New York he worked for five years in a variety of jobs, including errand boy at a commission house, hostler in a livery stable, telegraph operator, and finally as a longshoreman on the Brooklyn docks. Perhaps through his connections on the docks, Daly in 1861 managed to gain passage on a ship to Panama. From there, he traversed the malarial isthmus and then continued by land up the coast to San Francisco.

In California he worked for a time on ranches and farms but soon followed a friend into mining. For a while he tried his hand at placers, eventually gravitating toward work in established mines. By 1862, Daly landed at Nevada's mighty Comstock—the greatest silver mine in history and the largest, most sophisticated operation of its day.

It was at the Comstock that Daly truly learned the trade of mining: how to recognize fruitful veins; how to tunnel; how to timber; how to blast. His growing talent was rewarded with promotion to shift boss, and Daly would add to his skills the intangible quality of leadership. Throughout his career—in sharp contrast with his rival William Clark—Daly would be known as a miner's miner, a benevolent dictator beloved by his men.[16]

It was during his time in Nevada that Daly also forged important relationships—from his reporter friend Samuel Clemens (not yet writing under his later pen name of Mark Twain) to George Hearst, his eventual financier and partner. The West of the 1860s was a dynamic, booming place. Potential for a talented young man seemed boundless, and Daly made the most of every opportunity.

By the time Daly left the Comstock in 1868, he was widely recognized as one of the West's leading young miners. A new opportunity presented itself in 1870, when Daly's reputation drew the interest of the Walker brothers—four powerful merchant-bankers out of Salt Lake City. The Walkers hired Daly to run their Emma silver mine, then their Ophir. Both properties boomed. Around this time, the Walkers began to catch wind of opportunities in a Montana town called Butte. In 1876, they sent Marcus Daly to scout it out.

Daly, known for being able to "see farther into the ground than any other man,"[17] liked what he found. He settled on a recommendation that the Walker brothers purchase a silver mine called the Alice. The Walkers agreed, giving Daly an equity share to move to Butte and run the mine. The Alice—purchased for $25,000—would ultimately produce millions in silver for its owners. The mine would also launch Daly, and most significantly, signal the true takeoff of Butte's silver boom.[18]

Daly superintended the Alice for five years and in the process became a wealthy man. For Daly, though, the object was not wealth, but empire. In 1880, he sold his share in the Alice for "a rumored $100,000" and began searching Butte for another property.[19] What he found would surpass his wildest dreams, for nothing like it had ever existed before.

I n 1875, a former Union soldier named Michael Hickey staked a claim on the Butte Hill, hoping to catch the swelling wave of silver. As an infantryman in McClellan's Army of the Potomac, Hickey had fought in Richmond and Petersburg. Before the battle for Richmond, he read a Horace Greeley editorial in the *New York Tribune*. "Grant will encircle Lee's forces . . ." predicted Greeley, ". . . and crush them like a giant anaconda."[20]

The image of an "anaconda" would stick with Hickey, and when it came time to name his new claim, "That word struck me as a might good one."[21] In the five years following its establishment, the Anaconda was no better than a middling prospect in a town where scores of companies had staked their claims. But Marcus Daly turned the tide. According to Butte lore, Daly met Michael Hickey one day while taking a walk. Hickey told Daly about the progress at his mine. "I've sunk a forty-five-foot shaft on my Anaconda claim and I've sure got silver ore but I've got to go deeper to make it pay," he said. "If you'll deepen the shaft, we can make a deal."[22]

Daly would ultimately pay $30,000 for outright ownership of Hickey's property. He had taken the first step toward his future empire, but it was only the first step. As Daly knew better than anyone, to transform the Anaconda into a profitable silver mine would require money. Money to sink and timber the shaft. Money to excavate the ore. Money to crush. Money to smelt. Money far beyond his own not insubstantial means.[23]

There are conflicting stories as to why Daly did not stick with the Walker brothers of Salt Lake. By one account, he offered the Walkers an option on the Anaconda "but they could not be induced."[24] By another account "Daly froze them out." Even as he ran the Alice for the Walkers, Daly corresponded with his old friends from the Comstock—George Hearst and the partners in his syndicate. In 1881, Daly traveled to San Francisco to present his plans.[25]

George Hearst—father of the future newspaperman William Randolph Hearst—had reason to trust the ruddy Irishman from Butte. In his autobiography, the senior Hearst credited Daly with a tip to buy—for $30,000—a mine called the Ontario. The Ontario produced an estimated $14 million in silver and gold. "From that $30,000," wrote Hearst, "everything else came."[26]

Hearst and his partners (including Lloyd Tevis, who would eventually preside over the Wells Fargo Bank) bought into Daly's new Anaconda mine. Daly received 25 percent of the equity in the mine, management responsibility, and—most significantly—an almost limitless operating budget.[27] In June 1881, workmen began sinking a new main shaft at the Anaconda. They quickly found rich deposits of silver. In the early days of the mine, ironically, Daly leased a mill from his future rival—William Clark.[28] Clark, back from his study at the Columbia University School of Mines, was still frantically amassing his own early fortune.

The Anaconda was a profitable property on the basis of silver alone, but as Daly dug deeper, it was a less precious metal that would ultimately create a far greater treasure.

In late 1882, miners at the Anaconda were drilling at the 300-foot level when they began to encounter "new material." Word was dispatched to Marcus Daly and his lieutenant Mike Carroll, who came down to look for themselves. They watched as dynamite was set in a circle of holes bored into the quartz walls. The fuses were lit and the men took cover. A great blast shook the mine.

When the debris settled, Daley stepped forward to pick up a chunk of the smoking rock. He turned to Mike Carroll excitedly. "Mike," he said. "We've got it." It was a singular moment, and it would change the life of both Daly and his adopted city of Butte. Marcus Daly had discovered the largest deposit of copper in the world.[29]

Like many great men, Daly combined talent and hard work with good timing. At the Centennial Exposition in Philadelphia, just six years before Daly hit the Anaconda copper vein, a man named Alexander Graham Bell had demonstrated the transmission of speech through a copper wire. Two years before Daly's discovery, a man named Thomas Edison received a patent for an invention he called the "incandescent lamp." Through the inventions of Edison and Bell, the world was on the brink of an explosion in the demand for copper.[30]

Daly knew he had discovered the supply.

Four

"What Men Will Do"

*Ernest Sullau's Death an Example of What Men Will Do
to Help Their Fellows*
—HEADLINE, *BUTTE MINER,* JUNE 10, 1917

Ernest Sullau would have smelled gas within minutes of accidentally starting the fire, probably during those initial, panicked moments in the 2,400 Station. Sullau had encountered gas before. In fact, after his wife, Lena, learned of an earlier incident, she had urged him to be more careful. "Don't worry," he told her. "No death is easier and sweeter."[1]

Sullau's reassurances aside, death by gas falls far closer to insidious than sweet. This was particularly true for the men of the North Butte Mining Company. As they battled to escape from the depths in the early morning hours of Saturday, June 9, 1917, most of the miners knew that the deadly, invisible gas pursued them. They could smell it and even *taste* it as they ran.[2]

The gas was carbon monoxide, also known by its scientific symbol of "CO." According to the Bureau of Mines' official accident report, "At least 150 of the 163 men killed showed the cherry-red blood and knotted veins of neck and side of head characteristic of carbon monoxide poisoning."[3] CO is a brutally efficient killer. If inhaled in sufficient quantities, it keeps the blood from absorbing oxygen, strangling its victims from the inside out.

Carbon monoxide is produced when wood or fossil fuels are burned, and it is most dangerous when its fumes are confined. Thousands of people die each year from CO poisoning inside their homes, usually because furnaces or appliances fail to ventilate properly. It is easy to see how carbon monoxide could wreak havoc within the interior of a burning mine.

Most people familiar with carbon monoxide know that the gas is colorless, odorless, and tasteless. Yet dozens of North Butte survivors reported smelling and tasting gas, meaning that the CO must have mixed with other gases and smoke. One likely source of this distinctive odor was the burning, melting three-ton cable. Its cloth insulation was soaked in oil, and both the lead sheathing and copper-wire center would have spit off fumes as they melted in the intense heat of the fire. Another possible source of the odor was the burning shaft timbers, chemically treated to withstand moisture.

On the night the fire began, 415 men were at work underground in the North Butte's Granite Mountain and Speculator mines. They were widely dispersed, laboring in pairs or small clusters in hundreds of locations throughout the workings. Fifty-seven men were at work in the upper parts of the mines—between the 400- and 800-foot levels. Most of the rest worked deep in the ground—between 1,700 and 3,000 feet below the surface.[4]

Ernest Sullau grasped quickly that the fire put hundreds of men at risk, though he might not have imagined the scope or speed of the danger. As we know, the fire broke out at 11:45 P.M., driving Sullau and his companions from the 2,400 Station within minutes. By midnight, rising smoke was visible at the Granite Mountain collar.

Smoke and gas also spread rapidly from Granite Mountain across to the Speculator through the multiple crosscuts.[5] So rapidly did the fumes travel that by 12:10 A.M.—only twenty-five minutes after the start of the fire—smoke began to rise from the collar of the Spec. By 1:00 A.M., gases had spread to adjoining mines owned by the Anaconda Copper Mining Company, including the Badger, the Diamond, and the High Ore.[6]

As the miners initially encountered smoke, most thought it was dust from drilling or smoke from blasting, both commonplace in the mines. Indeed, the North Butte mines exploded more than 1,700 pounds of dynamite *every day*.[7] "My partner was at the machine when we noticed what appeared to us to have been an unusual amount of dust," said a survivor named Jack Watts. His partner started to oil the machine "when a boy from the 2,200-foot level rushed to [us] asking us if we saw any smoke. By this time we were all coughing."[8]

In many parts of the mine, the smoke and gas were accompanied by a wave of scorching heat from the inferno of the Granite Mountain shaft. Victims hundreds of feet from any flames were found with their faces singed.[9] So intense was the heat generated in the Granite Mountain shaft that it warped the steel cages in the Speculator shaft—800 feet away. According to one report, twelve men were killed in the Speculator when the heat from Granite Mountain snapped "clean" the cable to their cage. The cable was pulled to the surface, leaving the cage and its victims lodged at the 2,200-level of the shaft.[10]

What we know about the movements of Ernest Sullau after the start of the fire comes from stories told by survivors, many of whom owed him their lives. Having decided against the quick flight from the mine that would have saved his own life, Sullau set about warning his fellow miners of the calamity that he himself had unleashed.

By midnight, all electric lines to Granite Mountain had been burned out. Even if there had been power, Sullau knew that there was no alarm system in place to warn miners of a fire or any other danger.

In later years, Butte mines used the air supply itself as an alarm system. In the event of an emergency, a noxious odor was added to the oxygen that was pumped from the surface into the mines—similar to the scent added today to propane or natural gas. When miners smelled this odor, they knew to evacuate.[11] In 1917, though, the Butte miners could depend only on one another. Their safety hung by a fragile, human chain.

It appears most likely that Sullau initially climbed *down* from the 2,400 Station. His intention was probably to descend as low as possible, then work his way back up toward the surface, warning as many men as he could find.

One early sighting of Sullau comes from the 2,600. To move between levels, it is important to note, Sullau had to climb. The typical route would have been through a manway, climbing on a rickety wooden ladder. To move from the 2,400 level to the 2,600 was the equivalent of climbing down a twenty-story ladder (assuming ten feet as the equivalent of one story). For his only source of light, Sullau gripped the same carbide lantern that spawned the fire. And of course throughout all of his movements after the outbreak of the fire, Sullau traversed through smoke and increasingly dangerous levels of gas.

On the 2,600 level, approximately 600 feet from the Granite Mountain shaft, a young Eastern European emigrant named Mike Jovitich was eating supper with fifteen other men when he felt a strange sensation in his nose. Soon after came a cry of "fire" and the miners, "as if one man," ran in the direction of the Granite Mountain shaft. "I am young and had been in the mine only three weeks," said Jovitich. "So I followed like a sheep would."[12]

Dozens of miners made the same mistake that night—running *toward* Granite Mountain and the very maw of the fire, "whereas any other direction would have been safer."[13] Their choice of direction, though wrong, was not irrational. Just two months before, in April

1917, fire had broken out in an adjoining, Anaconda-owned mine called the Modoc. The Modoc fire was *still* burning on June 9. In fact, in two incidents since April, fumes from the Modoc had infiltrated the North Butte property. Both times, miners fled toward the Granite Mountain shaft, where they were lifted to safety. In the early morning hours of June 9, many miners probably believed that the smoke came from the Modoc and ran toward Granite Mountain—the same path that had led them to safety before.[14]

There was another reason for the men to seek safety via Granite Mountain. On the day the fire broke out (and apparently for a few days before), the Speculator shaft had been under repair.[15] The Spec suffered recurrent problems with shifting ground, which frequently pushed the shaft timbers inward to such a degree that the cage could not pass. When this happened, the shaft had to be closed temporarily while crews shaved back the timbers to create enough space for the cages.[16] In the emergency circumstances of the Granite Mountain fire, rescuers would press the Spec hoists into service to assist rescue efforts. But on the day of the fire, the men working in the Speculator were lowered via the Granite Mountain shaft.[17] Thus, it would have been natural for them to believe that their best exit—perhaps their only exit—lay through Granite Mountain.

Whatever the reasoning that led miners to run toward Granite Mountain, the consequences were usually deadly. According to the *Bureau of Mines Accident Report,* for example, it appears that "at least a dozen" men died on the 1,800 level when they ran toward Granite Mountain. Some of them probably ran directly past a connection to the adjoining Badger mine that would have led them to safety.[18] Most would have had no way of knowing.

Like the young Eastern European Mike Jovitich, eating his supper on the 2,600, many Butte miners were new to the job. The estimate of *monthly* turnover in the mines was a remarkable 20 to 45 percent.[19] In

1917, open positions were usually filled from the wave of Eastern Europeans who washed into Butte's train station each week, many coming directly from the Old Country after quick transit through Ellis Island. Most were illiterate and spoke little or no English.

Even if the men had been able to read, there were few signs in the mines to label directions. One of the recommendations of the accident report, seemingly basic, was that signboards "be erected to show direction to exits to other mines, to the manways which are in condition for travel, and to the main shafts."[20] The requirement for signage marking escape routes had actually been a part of state law since at least 1897, a fact the accident report fails to note.[21] It was only after the fire, according to one survivor, that "they placed signs around to tell men which way to go."[22]

As Mike Jovitich and his fifteen companions dashed toward the Granite Mountain shaft, they ran into Ernest Sullau, an encounter that saved their lives. Sullau stopped them and turned them around, telling them about the source of the fire and directing them toward the Speculator shaft. Jovitich and the others ran for the Spec, while Sullau continued his search for more miners to warn.[23]

Even with the benefit of running in the right direction, the Jovitich crew was swimming in gas by the time they made it to the 2,600 Station of the Speculator shaft. They climbed 200 feet up ladders to the 2,400 level "and were again halted by fire and gas." At some point they joined other men, all seeking to use the workings of the Speculator to make their escape. The group climbed another 200 feet. By the time they reached the 2,200 level there were twenty-six men, now led by a man named Jack Bronson. Bronson, a shift boss, was a friend of Ernest Sullau's.[24]

At the 2,200 level, forty stories above their starting point at the 2,600, Bronson told the exhausted men what they least wanted to hear—that they must keep climbing. Two men began to weep and said that they could not continue, but they rallied when the others urged them on. As a shift boss, Bronson would have known that there was a connection to the Badger mine at the 2,000 level of the Spec—a possible way out.

Soon, though, men began to fail. Some likely collapsed from the effects of the gas; some stumbled in the crosscut or tripped on rail tracks. At this point, some of the men were probably without their lights. Falling down could easily result in falling behind, and in the darkness, in their weakened state, falling behind could mean death. Six men were missing by the time the main group made the Badger shaft. They rang for the cage, which came immediately. "I never saw anything look so good as did the cage," said Jovitich.

Remarkably, though, the young man did not climb aboard. The station tender asked for a volunteer to go back and find the missing men, and it was Jovitich who led him back into the gas-filled workings. "The smoke was bad and after I walked 500 feet my knees went out and I fell."

Jovitich woke up in a hospital bed. "I wish I had been strong enough to save the others."

Those men not fortunate enough to encounter Ernest Sullau or another experienced miner had to depend entirely on their own instincts. At the 2,200 level, for example, a young man named W. T. Wynder led a small group that first ran toward the Granite Mountain shaft. Others in the group included Wynder's partner and "a Finlander" whose name is reported only as "Voko."[25]

As the group came closer and closer to the Granite Mountain shaft, they encountered a wall of smoke and fumes. In Wynder's description, "That cloud was black as night." They watched as men in front of them "dove right into that wave of poisonous fumes and every yard we went it became blacker and more terrible." Then the draft that pushed the cloud blew out their lanterns, casting the men into total darkness. They dropped to the floor, feeling for the rail tracks to find their way.

Wynder realized that they were moving in the wrong direction. "Boys, she's coming this way!" he yelled. "Turn back and we've got a chance!" Finally one of the men managed to light a match and Wynder was able to spark his carbide lantern. By this time, though, Wynder's partner was gone. Apparently disoriented in the darkness, he had crawled in the wrong direction and was later found dead.[26]

Others too began to panic. "I quit," said the Finlander Voko. "I cash in my checks." One of the other men punched Voko "to liven him up," and the remnants of Wynder's group—including the Finlander—fled back toward the Speculator. Most made it to safety.

E rnest Sullau was probably in the gas-filled workings of the mines for more than two hours. Though his precise path is unclear, it is estimated that the forty-eight-year-old covered more than three miles, "much of which was climbing up and down ladders." He spread his warning throughout the lower levels of the mine and sent "at least fifty men to safety."[27]

There are conflicting accounts of Sullau's push for the surface. The earliest account says he was leading a group of men through a connection in the Badger mine.[28] A later, more detailed account says he was leading men through the High Ore.[29] Whatever his precise location, both accounts agree on one central fact: Sullau had a final opportunity to flee the mine, but instead turned back.

He took this action fully aware of the potential consequences. According to a detailed story in the *Anaconda Standard,* three men grabbed Sullau and attempted to dissuade him from going back down, telling him it would mean death. But Sullau was determined to search for his friend Jack Bronson, the shift boss who earlier led other men (including the young Eastern European Mike Jovitich) up the long climb through the Speculator.[30] Bronson survived the fire, and it is possible that he was already safe on the surface when Sullau went back to look for him.

O n the 600 level of the Speculator, just after midnight, a debate took place that must have been repeated throughout the upper workings on the night of the fire: two men arguing about the source of smoke. One, a Balkan miner named Chris Vukovich, believed it came from nearby blasting. His partner, Louie Muller, worried that the cause was more ominous. "That's not powder smoke," he told Vukovich. "It's gas."

Then a shift boss ran through, settling the debate with frightening certainty. "Run boys, it's fire!"[31]

Vukovich and Muller were two of the fifty-seven men working that night in the upper portions (between the 400 and 800 levels) of the Granite Mountain and Speculator mines. Thirty-one of these men— more than one half—would die, an even higher mortality rate than the lower portions of the mines.[32] Though they were closer to the surface and farther from the source of the fire, the miners in the upper levels were in some ways disadvantaged compared to the men at the greater depths. The upper levels did not connect to adjoining properties—the most successful route to safety for the men below. The other disadvantage for the men in the upper reaches was that none of them had any idea as to the source of the fire, smoke, and gas.

As for Muller and Vukovich, both initially climbed down a narrow manway ladder toward the 700 level. They most likely chose this direction because the 600 level had no connection to the Granite Mountain shaft, whereas the 700 did. Like so many others, they probably believed that the Granite Mountain hoist would provide their passage to safety. As they descended, though, the two partners and the other men with them encountered thickening smoke,[33] pouring through the crosscut from the Granite Mountain shaft. "[T]he smoke and gas nearly suffocated us," reported Muller. "Somebody said to go back . . ."[34]

There was panic on the ladder, with some miners seeking to go back up even as others piled down. In the manway as elsewhere, the men struggled to sort through the chaos, their responses as diverse as the number of miners in the mine. Some resigned themselves to death, begging stronger men to pass along notes to their families. Some shouted curses. Some whispered prayers. Most struggled forward, desperate to breathe again in the light of day.[35]

Vukovich chose to keep climbing downward, and like almost every miner who went from the 600 to the 700 level, he was later found dead.[36]

Whether through knowledge, intuition, or luck, Muller was among the survivors. Unlike his partner, Muller turned around and climbed back up to the 600 level, deciding to try for the 600 Station of the Speculator. As he ran toward the station through the 600 crosscut, he stumbled across a pile of seven bodies. Muller managed to carry three to the station before becoming too weak to go back.[37] Someone rang for the Speculator hoist. Unlike the Granite Mountain shaft, where fire burned out all wiring, the Speculator shaft still had electricity. At least one of the Speculator hoists, though it had not been in use due to shaft repairs, was still serviceable.

A few minutes after they called to the surface, the hoist appeared. Muller would be one of the thirty-two men who were lifted from the 600 and 400 Stations of the Speculator shaft—the only men taken alive through either the Speculator or Granite Mountain shafts in the immediate aftermath of the fire.[38]

Two other men in the upper levels, unable to outrun the pursuing gas, showed the innovation that desperation bred. Their names were John Boyce and John Camitz. Like many others working in the upper levels, they made an initial effort to escape via the 700 level. Like a lucky few, they managed to retreat from the 700 before they were overcome.[39]

Instead of climbing up to the 600 level, however, Boyce and Camitz made their way into a wet drift, an excavation that branched off the 700 crosscut.[40] But they could not outrun the smoke and gas. The two men fell to the ground, unable to breathe. "We thought we surely must die," said Boyce later.[41]

"As I fell, half exhausted, to the ground, I felt the hose line carrying air." It was, literally, a lifeline. By 1917, Butte mines used compressed air to power their drills, and dozens of similar hoses were strung throughout the workings. Although nearly blind from the smoke and "almost out of my head," Boyce managed to cut two openings in the hose with his candlestick holder. "We lay there on the ground, our blouses pulled tight about our head, and sucked in that hose line air . . ."[42]

For nearly four hours they held this tenuous position, sucking air from the hose even as the gas washed over and around them. Several times the two men pulled themselves upright, creeping back toward the 700 crosscut to see if the gas had cleared. It had not, and they quickly retreated to their prostrate position with the hose.[43]

Around 5:00 A.M., events forced Boyce's and Camitz's hand: The air hose failed. With no other option, they began working their way toward the Speculator Station, stepping over the bodies of their fallen fellow-miners.[44] Many dead men were found with their blouses over their heads and their faces pressed to the ground, searching, it appeared, for that last breath of clear air.[45] Other dead men were found still gripping their lunch pails, overcome before they could grasp the full gravity of the danger that pursued them.[46]

As they approached the Speculator Station, Boyce and Camitz saw the beacon of a rescuer's light. The cage was called and the two men were whisked to the surface, "weakened by breathing gases, but able to walk at all times."[47]

Ernest Sullau almost made it. By the time he collapsed, he had cleared the property of the North Butte Mining Company and was ascending toward the surface through the adjoining Badger mine, leading a final group of miners to safety. For a while the other men dragged his limp body toward the surface, but finally they abandoned him, apparently afraid that their slower progress would cause them too to succumb to the gas.[48]

On the surface, a gasping miner reported Sullau's position to a growing group of rescuers. Fitted with primitive breathing apparatus, a crew descended the mine, located Sullau, and carried him to the surface. Though unconscious, Sullau was reportedly "still warm." The rescuers placed him in the Badger mine's "dry," a room where miners changed clothing at the end of their shift.[49]

A team of physicians, called to the mine in the minutes after the fire, launched a dramatic three-hour effort to save Sullau's life. They were well equipped, quickly connecting the stricken miner to a machine called a pulmotor. A recent invention, the pulmotor was a portable respirator turned by a hand crank. It looked a lot like another popular invention of the day—the Victrola.[50]

As many as fifteen doctors[51] labored over Sullau, working the pulmotor "in relays." At several junctures, Sullau gained consciousness, each time giving hope that his life might be saved. Ultimately, though, his body could not shed itself of the cumulative effects of the gas. Sometime around dawn on Saturday, June 9, the man who started the North Butte disaster was added to the list of its victims.[52]

As word of the fire spread through Butte in the predawn hours, panic-stricken families began to gather at the gates of the mines, held back by a company of troops from the Montana National Guard. One wife not among the worried bystanders was Lena Sullau. For three weeks, she had been in North Dakota, tending her ailing father.

"Saturday evening I received a wire from an undertaking establishment," Lena told a reporter upon her return to Butte. "This was the first news I had of the accident." She talked to the reporter about the effects of a recent tornado in eastern Montana, which she had witnessed on the train ride back to Butte. And she talked about the war in Europe. Ernest had an aged mother in Germany, but all of his other relatives had been killed in the fighting. "It seems all the world is wrong."[53]

Though several local newspapers lauded Sullau's bravery in warning other men of the fire, another, more insidious story was also spread. "Because the foreman had a German name," said Burton K. Wheeler, the federal district attorney in Butte, "it was widely believed [the fire] was an act of sabotage directed by the Kaiser."[54] The *Anaconda Standard* offered a similar report: "The suspicion is very strong in the minds of hundreds of people in this community that those interested in stopping the production of copper and zinc in this community may have had something to do with these fires."[55]

In the days to come, the rumor would mix easily in a deadly brew of anti-German hysteria, broader ethnic conflict, and a crippling strike. For the time being, though, all eyes remained fixed on the plight of the men in the mines.

In the first two hours after the fire, North Butte officials held out hope that most miners would escape through adjoining properties. Urgent telephone calls went out to the other mines as the officials attempted to establish a head count of those who had escaped. Sometime between 2:00 and 3:00 A.M., the timekeeper gave an initial report: 204 souls were still missing. "Scores of men," they suddenly knew, were "trapped in the lower workings."

Upon hearing the report, L. D. Frink, superintendent of the North Butte mines, turned solemnly to the other men in the room. The fire, he told them, looked "nothing short of a calamity."[56]

Five

·

"SWEETENED CORRUPTION"

By his example he has so excused and so sweetened corruption
that in Montana it no longer has an offensive smell.
—MARK TWAIN ON WILLIAM A. CLARK

By the mid-1890s, Butte, Montana, had become the undisputed copper capital of the world, and copper had turned Butte's two ruling "kings"—Marcus Daly and William Clark—into fabulously wealthy men.

For his part, Marcus Daly resided in his own town, christened with the same name—"Anaconda"—as his company. The town of Anaconda stood twenty-six miles from Butte and was built around a gigantic smelting operation constructed by Daly and his partners to process the raw riches ripped from their mines. Daly built and lived in the sumptuous Hotel Montana, which he kept fully staffed, though he and his family were often the only guests. Each morning, he ate a breakfast of beefsteak in a dining hall designed to accommodate 500, though he usually dined alone.[1]

The floor of Daly's hotel bar featured a wooden inlay of Tammany, his favorite racehorse, constructed by an imported New York artist from over a thousand pieces of hardwood. Anyone stepping on

Tammany's regal head was required to buy drinks for the house. As for Tammany himself, Daly kept him and the rest of his horses on a 22,000-acre horse farm in the lush Bitterroot Valley. (When Daly died, his horses sold at auction for more than $2 million.) For his commute between Anaconda and Butte, Daly rode a private rail car named Hattie, said to be the most luxurious in the country. Daly also owned the rails along which Hattie rolled, having built his own railroad after a dispute with Montana Union over rates.

William Clark was even richer—and more extravagant—than Marcus Daly. Clark's tastes, reflecting one sharp contrast with Daly, ran to the pretentious. He built, for example, a garish Fifth Avenue mansion in New York at a cost of $7 million, reportedly the most expensive private residence of its day. The mansion included 121 rooms, 31 baths, and 4 galleries for the display of Clark's beloved art collection, gathered during numerous trips to fin de siècle Europe. To ensure consistency of building materials throughout the house, Clark purchased entire stone quarries and his own bronze foundry. In addition to his Fifth Avenue residence, Clark maintained mansions in Butte and Los Angeles, with an oceanfront estate in Santa Barbara thrown in for good measure. Nor did Clark neglect his family's housing needs. Son William Junior's Butte home featured a $65,000 garage (with heated floors) for the protection and care of horses, carriages, and an extensive automobile collection.[2]

Daly's and Clark's financial power converted easily into political power, and it was the realm of politics that provided the central battleground for the two men's titanic clash. Speculation about the origin of the Clark-Daly feud has inspired a rash of theories ranging from personal slight to clashing business plans. Whatever the genesis of the enmity, the political ambition of William Clark and the election of 1888 provided the backdrop for the feud's first dramatic, public eruption.

In 1888, Montana was still a territory, and an election was held to select its nonvoting delegate to the United States Congress. Clark ran as a Democrat on a platform calling for lower trade tariffs (the Republicans of the day called for higher ones) and "keep[ing] the Mongolian race from our shores."[3]

Daly, in addition to his personal disdain for Clark, had a pointed parochial interest at stake. His copper industry required gargantuan amounts of timber—both to buttress his mine shafts and to fuel his smelters. Part of Daly's western empire included extensive timber holdings, but Anaconda also took logs another way—by poaching off federal lands. For years this practice had been overlooked, in part because of vague property lines, but the Democratic administration of Grover Cleveland brought several enforcement suits—still pending in 1888. Daly hoped that Republican Benjamin Harrison would win the presidential election (which he did), and that a Republican delegate from Montana would have more sway than a Democrat (i.e., Clark) in getting the Department of Interior to quash the suits.[4]

Daly quietly set about to engineer a Clark defeat, beginning by ensuring that his own miners and sawyers (whose shift bosses inspected the ballots before submission) voted against Clark. It worked. Clark lost fourteen of sixteen Montana counties, and the infamous "War of the Copper Kings" had begun.[5]

For a dozen years, the Clark-Daly feud would foul the waters of Montana politics—culminating with perhaps the most corrupt election in American history and spilling dramatically onto the floor of the United States Senate.

By 1899, Montana had become a state—entitled to representation by two senators in Washington. Until 1913, when the Seventeenth Amendment to the U.S. Constitution came into force, senators were not elected directly by popular vote, but rather indirectly by state legislatures. This concentration of electors greatly facilitated corruption, conveniently congregating the handful of men who cast the deciding ballots.

The stage was set for copper king William A. Clark, who having conquered the world of business now ached for the title of senator. He made no bones about the means he would deploy to win. His son Charlie said, infamously, "We will send the old man to the Senate or the poorhouse."[6]

An investigation by the United States Senate would later reveal the depth and breadth of Clark's bribery. For many legislators, the bribe was no more complicated than an envelope stuffed with cash. For others, bogus business deals were concocted to cover the tracks, including a number of land transactions in which Clark paid ridiculously inflated prices. One legislator struggled to explain how he had arrived at the legislative session penniless (indeed, borrowing $25 to make the trip) but returned home to pay $3,500 cash for a new ranch. Another claimed part of his $3,600 windfall as the profits of gambling; the rest, he said, he had "found in his hotel room."[7] By one credible estimate, Clark paid around $431,000 for the purchase of forty-seven votes, a tidy sum in the currency of the day.[8]

This scale of bribery, to the credit of a few solid Montanans, did not go unnoticed. Indeed, a remarkable drama played out over a month-long period in Helena, the state capital. Rumors of Clark's bribery began even before the opening of the legislative session, and during the session, "the purchase of votes was talked about almost as freely as the weather."[9] Mark Twain, who knew William Clark personally, said that "by his example he has so excused and so sweetened corruption that in Montana it no longer has an offensive smell."[10]

The smell, though, *was* offensive to some—or at least to a few. Rumors became so persistent that the Montana legislature formed a committee to investigate the corruption before the voting took place. The most dramatic testimony came from a former buffalo hunter,

Senator Fred Whiteside. Whiteside reported that he had been offered a bribe by Clark agents, which he accepted in order to provide evidence. Dramatically, Whiteside presented $30,000 in cash to the investigating committee. "I know that the course I have pursued will not be popular, but so long as I live, I propose to fight the men who have placed the withering curse of bribery upon this state." A grand jury was now convened to consider the findings.[11]

William Clark, though, was far from finished. As balloting began for the Senate election and as the grand jury convened, Clark's men flooded still more money into the streets—now including the grand jurors on their lists of targets. Clark's network of Montana newspapers, meanwhile, blasted the Whiteside allegations as part of a vast conspiracy.

At the same time, Senator Whiteside himself became the target of a vicious counterassault. Whiteside's recent election to the state senate had been challenged by his local opponent, and Clark's forces seized upon the opening. A recount was demanded, and Clark's forces succeeded in disqualifying ballots in which Whiteside voters had marked an "X" *after* his name—rather than *before*.[12]

Clark's audacious, shameless counterassault worked. The grand jury declined to find evidence of bribery. Senator Whiteside, meanwhile, was disqualified from the state congress. The old buffalo hunter, at least, did not go down quietly. On the day of his disqualification, Whiteside rose on the floor of the state senate and gave his former colleagues full bore. "If I failed to express myself at this time," he said, "I feel that I would be false to myself, false to my home, and false to the friends that have stood so manfully by me.

"Let us clink glasses and drink to crime," chided Whiteside. The Senate election, he said, "has reminded me of a horde of hungry, skinny, long-tailed rats around a big cheese." Dozens of men would turn away their eyes and squirm uncomfortably in their seats as Whiteside delivered his final broadside. "I am not surprised that the gentlemen who have changed their votes to Clark recently should make speeches of explanation, but I would suggest that their explanations would be much more clear and to the point if they would just get up and tell us the price and sit down."[13]

Money, indeed, had trumped. The Montana legislature sent William A. Clark to the United States Senate, but the election saga was far from over.

Marcus Daly's newspapers had made sure that the bribery scandal received wide play. Organized in part by Daly, anti-Clark forces took their fight to Washington. The U.S. Senate, under its own rules, can reject members for cause. On the day that William Clark took his seat in the Senate, two petitions were laid before the national body—one from Clark opponents in the Montana legislature and one from the governor of Montana. The petitions outlined the charges against Clark, and in a process familiar to modern political observers, the Senate launched an ethics investigation through the Committee on Privileges and Elections.

Clark watched as a parade of Montana state legislators gave sheepish explanations of the sudden turns in their economic fortunes. Senator Whiteside too told his story, and both Clark and Daly were called before the committee. (Daly, of course, was no angel. The committee, for the record, found ample wrongdoing on both sides.)[14]

In the end, Republicans and Democrats voted unanimously "that the election to the Senate of William A. Clark, of Montana, is null and void on account of briberies, attempted briberies, and corrupt practices." Faced with adoption of the committee recommendation by the full United States Senate, William Clark resigned on May 15, 1900.[15]

Resignation, though, was not the same as quitting. Remarkably, Clark's most outrageous maneuvers were still to come. His resignation created a vacancy that the governor of Montana would now be entitled to fill. The governor, Robert B. Smith, was a Daly ally and had aided the effort to overturn Clark's election. Smith's *lieutenant* governor, however, a man named A. E. Spriggs, was a close ally of Clark's.

Clark allies arranged an elaborate scheme to lure Governor Smith out of the state. In his absence, Lieutenant Governor Spriggs—now *Acting Governor* Spriggs—appointed William Clark to the vacant Senate seat! Learning of this outrage, Governor Smith rushed back to Montana, declared Spriggs's action invalid, and appointed his own man to the Senate. Each side protested the action of the other, and both candidates were tossed to Washington for the United States Senate to sort out.

Perhaps fatigued with the shenanigans, the Senate never resolved the matter of Montana's rightful representation. The issue would die without *any* man taking the office. Montana citizens, having labored to earn the cherished right of statehood only a decade earlier, would have only one senator in Washington for the balance of the term.[16]

T he election scandal of 1899, viewed from the distance of a full century, might be amusing if not for the legacy it spawned. Through a decade-long struggle founded on ego and personal aggrandizement, the Copper Kings had done much to "define deviancy down."[17] In the process, Montana picked up habits that would stick with the state for decades, including crude manipulation of the press, naked political corruption, and domination of state government by the copper industry.

In a few years, powerful interests from outside the state would use these same tools in ways as yet unimagined. Indeed, in one sense, the War of the Copper Kings can be viewed as a "noisy diversion."[18] For while Clark and Daly roiled the waters on the surface, a great shark circled below.

Six

"HELMET MEN BRAVING DEATH"

Helmet Men Braving Death Each Minute
—HEADLINE, *BUTTE DAILY POST,* JUNE 9, 1917

Cornelius "Con" O'Neill was lying in bed when he heard the Speculator whistle blow the alarm, just after midnight on June 9, 1917. O'Neill was foreman of the Bell and Diamond mines, Anaconda-owned properties adjoining the Speculator to the south. Married and a father of four young children, the thirty-seven-year-old lived with his family in a gracious house (owned by Anaconda) barely a stone's throw from the mines.[1] O'Neill's wife, Julia, sensing the seriousness of the emerging disaster, pleaded with him to stay home. Instead, the "big, robust Irishman" rushed to the Bell-Diamond.[2]

O'Neill arrived while the fire was still in its early minutes, probably around 12:30 A.M. He found himself in the midst of a chaotic and confusing scene. What was obvious, though, was that the Granite Mountain shaft was ablaze. The "flaming torch" at its collar stood only a few hundred feet up the hill. Foreman O'Neill instantly recognized the danger to his own men in the depths of the Diamond. Of particular concern were thirty miners at work on the 1,800-foot level, where there was a direct connection to the Speculator.[3]

O'Neill directed his men to gather the makings for a bulkhead to "keep the smoke and gas from our men." Quickly they loaded canvas and other materials onto the cage. Two other men accompanied Con O'Neill into the Diamond shaft—Ed Lorry and "Con" Toomey (O'Neill was sometimes known as "Big Con" to distinguish him from the multitude of "Cons" among the heavily Irish miners). None of the men wore any type of breathing apparatus, but when they reached the 1,800 Station of the Diamond, the air initially was clear. They progressed some 150 feet toward the Speculator and began to make preparations to build a canvas bulkhead.[4]

When the gas came, it hit hard and fast. Lorry suddenly collapsed. O'Neill and Toomey attempted to carry the stricken man back to the 1,800 Station, but then O'Neill too went down. Though they were close to the station, only Toomey managed to crawl back. By the time he got there, rescuers in breathing apparatus had already descended to search for the three men, worried by the length of time they'd been gone.[5]

"To hell with me—I can make it," Toomey told the rescuers. "O'Neill and Lorry are in there." The rescuers hurried down the drift, finding O'Neill and Lorry unconscious. One of the rescuers described the scene. "The smoke and gas were so thick in the 1,800 of the Diamond that we could not see three feet in front of us. No man could live in that gas without a helmet for ten seconds." He was right. The rescuers managed to pull O'Neill and Lorry to the surface, but it was too late.[6]

The death of Con O'Neill, "the best known of the men who lost their lives in the disaster," contributed to the shock on that first day after the fire. His photo in the newspaper, showing a powerful man with a thick mustache and thicker neck, underscored the vulnerability of all the men. The Butte papers lauded O'Neill in the gushing tone of a less cynical age: "[H]is efforts to save his companion," wrote the *Butte Daily Post*, "even when he knew his own life was in danger, were characteristic of the man."[7]

Like the effort by Con O'Neill, other early rescue attempts were largely spontaneous, more a function of impulse than contingency plans. The main body of trained rescuers would not arrive at the mines until after dawn.[8] Instead of waiting, early rescuers simply dived in. Some, like Con, took their chances without the breathing equipment that would ultimately play a vital role in rescue efforts.

By the early twentieth century, breathing equipment was widely used by both mine rescuers and firemen. In the trenches of Europe, the apparatus was also used in connection with the "recent introduction of asphyxiating gases for offensive military purposes." Specifically, the helmets were worn by soldiers "setting off the gas."[9]

Rescuers at Butte used two main types of breathing apparatus: the German-made "Draeger" and English-made "Fluess." Both the Draeger and the Fluess looked similar to deep-sea diving equipment, complete with breathing helmets. It was the helmets that gave the mine rescuers their popular name—the "helmet men." In addition to the helmet (or in some models, a skullcap with goggles and mouthpiece), the apparatus consisted of compressed oxygen tanks that were worn on the back and a rubber "breathing bag" that was worn on the chest. Inside the breathing bag was alkali to absorb the carbon dioxide exhaled by the user. Various tubes and valves connected the components. Altogether, the helmet men's breathing equipment tipped the scales at a hefty forty pounds. It was hot and cumbersome, particularly in the sweltering depths of the mines.

The breathing apparatus of the day was also notoriously unreliable. Indeed, a 1917 report on breathing equipment by the U.S. Bureau of Mines concluded that its use in *any* circumstances "involves grave danger." Citing the many deaths of rescuers wearing the equipment, the report called for a fundamental redesign. The authors noted that helmets were prone to leaking, subjecting the wearers to gas.

The 1917 equipment was also incapable of *increasing* the amount of oxygen made available to users during strenuous activity. Obviously, rescuers frequently faced situations demanding heightened activity. The report used the specific example of rescuers carrying injured men to safety, "an exertion of which men hampered with helmets are practically incapable."[10] Helmet men who overexerted themselves frequently passed out—and sometimes died.

Despite its limitations, the breathing apparatus was all that Butte's rescuers had. In eight days of rescue and recovery work, an estimated 175 men used the breathing apparatus to make more than 2,000 descents into the burning mines.[11] Remarkably, though many were injured, none died.

The first rescue effort with breathing apparatus occurred barely an hour after the fire started. Sometime before 1:00 A.M., two men, named Harry Goodell and William Burns, entered the mine, using the Speculator shaft to access the 700 Station.[12] They recovered two bodies, carrying them to the surface at 1:30 A.M.

The name of the first identified victim was Bill Ramsey, a mucker who worked on the 700 level of the Spec.[13] Muckers occupied a position near the bottom rung of the mining ladder, their job to "muck up" the ore that more skilled workers had dynamited from the walls. Ramsey had been busy shoveling rock when smoke from Granite Mountain began to fill the drift.

Panic broke out, with many men on the 700 scrambling to climb *up* a manway to the 600 level even as men on the 600 tried to climb *down*. One of Ramsey's fellow miners, John "Shorty" Thomas, urged him to climb up to the 600, telling him that there was no escape at the 700 level. But Ramsey refused. Apparently there was a brief argument about which way to go. Later Thomas, safe above ground, would report Bill Ramsey's last words: "I'm a stayin' fool."[14]

It wasn't until 2:15 A.M. on June 9, when North Butte officials first realized how few men had escaped through adjoining mines, that the organized rescue effort began to take full shape. The officials gave immediate notice to the coroner and the hospitals. The coroner, Aeneas Lane, called the undertakers. Within twenty minutes, a small fleet of ambulances and "undertakers wagons" stood waiting inside the mine yard.[15] They would be busy throughout the long night.

Two lowly timekeepers, Grover McDonald and Thomas O'Keefe, emerged as key players in the early efforts to respond to the disaster. Their prime asset was a knowledge of the names and faces in the mines, and to them fell the task of keeping a tally of who was dead, alive, or missing. An urgent call went out, ultimately picked up by the newspapers, for North Butte miners to report in.[16] Many men who had emerged safely from the fiery mines hurried home so that loved ones would know they were alive. The *Anaconda Standard* painted a telling portrait of a group of miners brought up through the Badger. They "took one look around the mine, glanced toward the Speculator and then started at a brisk walk for town." Once on the street, a few threw their lunch buckets in the ditch. "I'm through with mining from this time on," declared one man. "I will never go into the hole again."[17]

The call for rescuers extended, of course, to all available helmet men. The federal government provided one source of trained men and equipment. Daniel Harrington, a Butte representative of the Bureau of Mines, sent telegraphs requesting the dispatch of two specially equipped mine rescue cars. Though much ballyhooed by Harrington, who would ultimately author the official Bureau of Mines report on the accident, the cars did not play a vital role in the rescue effort. The nearest of the two was located in Red Lodge, Montana—250 miles away. The second was in Colorado. By the time both cars reached Butte, the mission had shifted from the rescue of the living to the recovery of the dead.[18]

The more significant source of helmet men and equipment was local—the so-called Safety-First crews maintained by all of the major mines. As the largest mining operation, Anaconda provided the largest group of Safety-First men, more than 60 percent of the 175 men who wore equipment during the disaster. Though volunteers, the helmet men were paid time and a half for their work, or $7.12 per shift.

By noon of June 9, twelve hours after the start of the fire, thirty helmet men were at work below ground. Thereafter, more than fifty helmet men were deployed during any given eight-hour shift.[19]

Ninety of the helmet men had received formal training by the Bureau of Mines. One goal of the training was to teach the men to avoid the types of physical exertion with which their breathing equipment could not keep pace. Thus they were taught to walk at a speed of only three miles an hour, resting at the end of each minute. The men also practiced in dark "smoke rooms." Wearing their breathing apparatus, they engaged in activities such as carrying a dummy, crawling through a nineteen-by-seventeen-inch tunnel, and climbing over obstructions. The exercises were important and succeeded in simulating many of the physical challenges that the rescuers would face. But as the Bureau of Mines noted, "It is not possible to reproduce the mental strain, anxiety, and fear sometimes manifested in mine-recovery work."[20]

In addition to the helmet men, mine officials, firemen, medical teams, and undertakers—even the U.S. Army—were pressed into service. Company F of the Montana National Guard happened to be camped in Butte on the night of the fire. The Montana Guard had been federalized after the U.S. entry into the war, and Company F was stationed in Butte on "utility duty"—guarding the mines against potential sabotage. Only three days earlier, the soldiers had put down the draft riot. On the night of the fire, they formed a cordon around the mines, keeping out everyone except rescuers, doctors, and undertakers.[21]

Butte mayor William Maloney was the last major figure to join the rescue forces, awoken from his bed at 3:00 A.M. and rushed to the North Butte by car.

M any citizens of Butte first learned about the North Butte fire when they woke up on Saturday morning to find the town blanketed in smoke. The smoke "poured in a torrent, deluging the valley below and spreading out over the side of the hill like a giant shroud."[22]

From the time, minutes after the start of the fire, when an unnamed miner attempted to throw a ten-gallon bucket of water down the Granite Mountain shaft, efforts to battle the flames and fumes were continuous.

For the first five hours of the disaster, rescue officials deliberately allowed the Granite Mountain shaft to burn. Though the draft in the Granite Mountain shaft was normally downcast, the flames and heat caused it to turn upward. Officials hoped that this upcast draft would help to suck the smoke and gas from the rest of the workings.[23]

Rescuers took several steps in an effort to augment the updraft effect. At 12:15 A.M., only thirty minutes after the start of the fire, North Butte General Manager Norman Braley ordered a reversal in the direction of a gigantic fan at the mouth of the Rainbow shaft. It was normally a suction fan, but Braley hoped that blowing *down* the Rainbow would help to push fumes *up* Granite Mountain. At 1:00 A.M., the same step was taken with fans at the Speculator and Gem shafts. And over the next forty-eight hours, several more mobile fans would be set up—including one so large that it was pulled into position by "eight horses and 50 men."[24]

The combined effect of the fans was "a cloud that blew out of the shaft like it was driven by a gale."[25] The overall effect of these efforts seems to have been mostly positive, contributing to the more rapid clearing in some sections of the mines.[26] While this clearing did not occur rapidly enough to save the 163 victims of the fire, it would later become critically important to dozens of other men.[27]

The fans also allowed firemen to begin dousing the blaze with water. Initially, rescuers had worried that dumping water into Granite Mountain might have the effect of turning the shaft back into a downdraft, which would in turn have sent more smoke and fumes into the workings. With the augmentation of the draft by multiple fans, however, firemen believed that they could begin to fight the fire with water. Eventually, "four pipe lines" would pump millions of gallons of water into the shaft at a rate of 500 gallons per minute. Still, it would be five days before firemen considered the fire to be under control.[28]

The water did have some unintended consequences. It contributed, along with the fire's incineration of shaft timbers, to the cave-in of large portions of the Granite Mountain shaft.[29] Another consequence was more gruesome. Helmet men found a number of victims who had apparently been scalded to death. They had been too near to Granite Mountain stations when the water was dropped. The great heat of the shaft fire instantly vaporized the liquid, creating a wave of deadly steam.[30]

John L. Boardman was a slight man with a weighty responsibility. As the head of Anaconda's Safety-First crews, Boardman commanded the growing army of helmet men who began descending the mines in the early morning hours after the fire. Boardman had an additional layer of accountability, having personally trained many of Butte's helmet men.

Photos of the rescuers taken before the disaster show men who look more like fighter pilots than miners. They model their technologically advanced equipment with an air of pride and even swagger, though a leavening of maturity is also apparent. In an era when ever-growing percentages of miners were unskilled, the Safety-First crews were drawn from the ranks of the most experienced men.

J. L. Boardman arrived at the North Butte properties shortly after the start of the fire and began an eight-day saga that would test his men's experience. His style of supervision was described as "calm, cool, and collected." He personally inspected each breathing apparatus before fitting it to a rescuer's head. "Look to the valves on the helmets and take every precaution," Boardman told them. "You must look to your own lives."[31]

In contrast to the earliest rescue efforts, Boardman helped to organize a systematic sweep of the mines. With the Granite Mountain shaft spewing flames, he channeled the helmet men through adjoining mines—especially the Speculator, the High Ore, and the Badger. Within each mine, the helmet men descended systematically, "one level at a time," beginning at the stations and then expanding their searches outward into the twisted braids of the workings. Because many rescuers were unfamiliar with the North Butte workings (especially those from Anaconda crews), miner-guides sometimes volunteered to lead the way.[32]

They faced a pitch-black maze encompassing hundreds of miles of underground passageways—most filled with smoke that was "saturated with steam"[33] and laced with deadly gas. Each level was its own intricate maze. There were few straight lines beyond the crosscuts, the drifts instead following the haphazard path of the copper veins. For their only light, the helmet men carried "Ever Ready" flashlights. When the beams could not penetrate the murk, the rescuers dropped to their knees and followed rail lines, an action impeded in some places by the knee-deep ash on the floor.[34]

Obstacles frequently blocked clear passage through the narrow crosscuts and drifts. In 1917, Butte mines still used a mixture of horses and electric engines to pull ore-laden "trains." The horses, like the miners, attempted to flee the fire and fumes. Dozens of animals lay dead along the tracks, sometimes with ore cars piled up behind them.[35] The low ceilings forced the helmet men to pick their way over such obstacles in a flat-out position, all the while taking care not to bump the helmets on their heads, the breathing bags on their bellies, or the oxygen tanks on their backs. (And, of course, all the while seeking to avoid physical exertion!)

Air temperatures in the depths of the North Butte ranged from the low eighties to more than a hundred degrees, with humidity between 85 and 100 percent. The breathing apparatus, in addition to being extremely hot and cumbersome, raised the temperature of the air that the men inhaled by at least an additional ten degrees. The Fluess model sometimes raised the temperature of inhaled air to 140 degrees. As the 1917 study on apparatus noted: "[A]fter an explosion and fire there is often a high temperature, and the air is saturated with moisture. These conditions the apparatus can not be expected to overcome."[36]

Yet the helmet men forged ahead. By the time they returned to the surface, their bodies were soaked in perspiration, their clothing reeked of gas, and their eyes were bloodshot from the smoke and fumes.[37]

For some it was worse. Many rescuers suffered injury when their helmets malfunctioned, including at least twenty "part or total prostrations."[38] Charles Fredericks, one of the helmet men who pulled Ernest Sullau from the mine, remembered detecting the "sweet odor of gas." He ended up unconscious for several hours, pulled to the surface by his fellow rescuers and ultimately hospitalized.[39]

Another helmet man experiencing equipment problems was a Speculator shift boss named William Budelière. He noticed that his helmet was leaking while exploring the 1,800 level with four other men on the morning of June 9. He informed the others of his plight before a second man noticed that his helmet also was leaking. They hurried to the station, rang for the cage, and managed to make it safely to the surface. Within thirty minutes, Budelière was reequipped and back underground.[40]

As a shift boss at the Speculator, William Budelière had an intimate knowledge of the North Butte workings. Both his knowledge of the mines and his fortitude would soon be put to the test. Barely three hours after the malfunction of his helmet, Budelière and twelve comrades were on yet another foray into the mines—this time the first rescuers to explore the lower levels of the Speculator and Granite Mountain shafts. Miners from as low as the 2,400 level had managed to escape through adjoining mines, and there was hope that other men might still be alive below ground.

After an initial descent from the Rainbow shaft, Budelière and the others worked their way "through miles of poisonous fumes" to the 2,000 level of Granite Mountain, only to encounter a new enemy—water.[41]

At the lower levels of the workings, mining companies contended constantly with groundwater. Miners dealt with water problems the same way that homeowners deal with leaky basements—sump pumps. To keep the deeper workings clear of water, giant electric pumps worked twenty-four hours a day. The Granite Mountain fire, though, had cut electricity. With the electricity out, the pumps shut down, and the groundwater began to rise—augmented by the great quantities dumped down the shaft by firemen.

In the crosscut between the Speculator and Granite Mountain, Budelière and his comrades encountered water up to their armpits—rising water. Because of the crosscut's low ceiling, only a few inches separated the surface of the deep water from the rocky roof of the tunnel. Budelière elected to keep going. For the other men, though, it was too much, and they decided to turn back.

As he slogged through the deep water, holding his head and his flashlight in the narrow band of air, Budelière reached a point where the rock above him was so low that his helmet scraped the ceiling. Incredibly, he removed the helmet and moved forward, craning his head in the narrow belt of air between the water and rock.

Budelière became the first (and for a while, the only) rescuer to reach the lower levels of the Granite Mountain workings. A *Butte Miner* headline hailed his courage as "Only Equaled by Those of *Titanic* Disaster." He found no survivors, instead only "scores of bodies."

Dozens of other helmet men would also play vital roles in those first hours after the fire, "assisting the almost exhausted men" who had clawed their way close to the surface. The general chaos made it difficult to estimate the precise number, though as the accident report states, the helmet men "undoubtedly saved many lives."[42]

A n increasingly well-organized system greeted the helmet men and the wounded miners as they were raised from the depths. At the Speculator, the Badger, and the High Ore, rough mine buildings were quickly converted into field hospitals and morgues.[43]

At the Speculator collar, a guard monitored entrance and exit from the shaft. Still, miners and rescuers rushed forward hopefully whenever the hoist was raised. As each man emerged (or was carried) from the cage, timekeepers O'Keefe and McDonald attempted to make an identification. Many miners, though conscious, clearly suffered from the effects of gas. In its early stages, exposure to carbon monoxide can cause confusion, irritability, and delirium. Miners were described as "muttering unintelligibly to all questions of the hospital attendants."[44]

A doctor with a stethoscope checked the unconscious, "determining as fast as possible the chance of a spark of life."[45] Those with a heartbeat were usually hooked up to pulmotors and, like the ill-fated Ernest Sullau, attended by a team of physicians. Many recovered and walked home, while ambulances transported others to local hospitals.

For the first thirty-six hours after the start of the fire, the dead were carried into the Speculator storeroom (and its counterparts at the other mines). Bodies fell into two categories—those that could be identified and those that could not. The unidentified were laid out in a row until friends could come along and attach a name. Other unidentified bodies were distributed among Butte's many funeral homes. Dozens of men would ultimately be buried, anonymous, in communal graves.

A t 3:45 in the afternoon on Saturday, June 9—a little more than twelve hours after the start of the fire—Coroner Lane issued a statement to the press, grimly tolling the growing carnage. Thirty-three bodies had been recovered from the mine, with ten of those still unidentified. Between 165 and 175 men were believed to be still underground. The statement's conclusion was grave: "[L]ittle or no hope is held out for them."[46]

In the depths of the mine, though, hope was still alive.

Seven

•

"STANDARD OIL COFFINS"

They will cut your wages and raise the tariff in the company stores
on every bite you eat, and every rag you wear.
They will force you to dwell in Standard Oil houses while you live,
and they will bury you in Standard Oil coffins when you die.

—F. AUGUSTUS HEINZE, OCTOBER 26, 1903

O n April 27, 1899, predawn of the twentieth century, fifty-seven-year-old Marcus Daly sold his beloved Anaconda to the Standard Oil Company.[1] Less than twenty years had passed since Daly, the once penniless Irish immigrant, bought a middling silver prospect for $30,000 from a veteran of the Civil War. The Anaconda that Daly sold to Standard Oil's William Rockefeller and Henry Rogers—for a stunning $39 million—was now a sprawling empire. The transaction, gushed the front page of the *New York Times,* was the "biggest financial deal of the age."[2]

The deal included not only Daly's rich Butte copper holdings, but also the other manifold components of Anaconda's vertically integrated machine: the world's largest smelting operation, timber holdings of more than a million acres, sawmill facilities in both Montana and Idaho, coal beds in Montana and Wyoming, waterworks, rail lines, a real estate company, brickyards, newspapers, hotels, and even retail stores.[3] Taken together, a remarkable *three-quarters* of Montana wage earners drew their checks from some part of Daly's Anaconda enterprises.[4]

Yet just as Marcus Daly had dreamed of creating an empire, Standard's Rockefeller and Rogers envisioned something bigger still: not *an* empire, but *the* empire. In the late nineteenth century, the mighty Standard Oil had come to dominate 90 percent of the American oil industry. Seeing the enormous profits to be gained from capturing a monopoly, businessmen of the 1890s battled to replicate the Standard Oil model in other fields—from sugar and tobacco to telephones and steel.[5] A new generation of Standard Oil leaders, meanwhile, searched for fresh domains to conquer. Copper stood out as a gleaming target. "Think it over," Henry Rogers is reported to have said, "and you will agree with me that the possibilities are far beyond those of oil."[6]

The purchase of Anaconda from Marcus Daly represented the first significant step toward the creation of a copper trust. Standard Oil incorporated a new holding company called "Amalgamated," into which the component pieces of the new copper trust would be collected. Henry Rogers—known to his Wall Street contemporaries as the "Hell Hound"—would be Amalgamated's new president. William Rockefeller—son of the great founder—would be secretary and treasurer. Marcus Daly, though nominally Amalgamated's vice president, was by this time gravely ill and increasingly unable to work. Eastern management had come to Butte.

So had eastern finance. Within a matter of weeks, Standard Oil took Amalgamated public. After a heavy promotional campaign to "stir up a market," Standard Oil sold one-third of the stock in its $39 million investment for a tidy $26 million in cash.[7] Put another way, Standard recouped two-thirds of its month-old investment in exchange for only one-third of Anaconda's ownership. In his classic history *The War of the Copper Kings,* C. B. Glasscock summarized the new situation back in Butte:

No longer were the mines of Butte to be operated solely for their
production of copper, zinc, silver, lead and gold. Thenceforth they
were to be operated also for their production of new stock
certificates, theoretically based upon metal resources and output,
but practically based upon the acquisitive genius of Henry H.
Rodgers [sic], William Rockefeller . . . and their associates and
successors in Wall Street.[8]

Standard Oil floated a second issue of Anaconda stock in 1901.
Naked manipulation of the stock price would follow: driving the price
higher before selling; then driving it down before buying back shares
at bargain-basement prices. Thousands of small investors lost millions of dollars.[9] Meanwhile Standard Oil/Amalgamated—with a fat
new war chest—went after the rest of Butte's copper.

Two men stood in its way. One was the scandalous William A.
Clark—fresh from his foiled effort to bribe his way into the U.S. Senate. The other was a relative newcomer to Butte, a twenty-nine-year-
old boy-king who would rise to fight a brilliant war against the most
powerful economic juggernaut of its time.

Butte history has a singular ability to keep topping itself. As remarkable as was Clark's and Daly's War of the Copper Kings in the
1890s, it practically pales in comparison to the titanic battles that took
place from 1899 to 1906. This second War of the Copper Kings would
again seize the attention of the entire nation. For decades to come, the
outcome of the conflict would transform the town of Butte, the state
of Montana, and the lives of the men and women who lived there.

The third of the three Copper Kings—Frederick Augustus
Heinze—was born in Brooklyn on December 5, 1869. Unlike
his counterparts, Marcus Daly and William Clark, young "Fritz"

Heinze grew up in a comfortable life of moderate wealth. He received a top-flight European education that included study in the classics, language, and a solid foundation in the scientific discipline that would shape his meteoric life—geology.[10]

Fritz Heinze entered the Columbia University School of Mines at the age of fifteen and received his engineer of mines degree at nineteen. In addition to a knowledge of mining, Columbia taught Fritz other skills that would serve him well in Butte. He joined the Phi Delta Theta fraternity and was known as an expert poker player. For the rest of his short life, Heinze would be the charming frat-boy rogue. He could drink with the best of Butte—and that was saying something. He was tall, physically powerful, and handsome—his giant ego buffered by raw charm. In short, people loved him.

After Fritz's graduation from Columbia, his father offered to fund postgraduate study in Germany. Fritz, though, had his heart set on the West. Three months later, in September 1889, Heinze arrived in Butte. He hired on as a junior engineer at the Boston & Montana Company—an independent operation owned by neither Clark nor Daly.

For a year the young Heinze added practical experience to his formal education. During the day he conducted surveys of the Butte Hill and at night he studied the complex veining of the richest hill on earth. He quickly became a master of Butte's hidden treasures, and what he learned convinced him there was still a fortune to be made—even as the titans Clark and Daly seemed to tighten their grip on the town.[11]

By 1892, the twenty-two-year-old Fritz Heinze was ready to put his theory to the test. He used a combination of family and outside money to start a new company—the Montana Ore Purchasing Company—"as a mining, ore purchasing, and smelting concern."[12]

An early Heinze business transaction reveals much about the tactics and ethics that he would bring to his career in Butte. Heinze leased the right to take ore from a Butte mine called the Estella. Under the terms of the lease, Heinze promised to pay the owner a handsome 50 percent royalty on all the ore extracted that showed at least 15 percent copper content. For any ore below 15 percent content, however, Heinze would pay nothing. As the lease played out, it turned out that *none* of the ore exceeded the 15 percent threshold—this in a mine known for rich deposits. How could this be? Because Heinze directed his miners to mix the good ore with waste rock, insuring the trigger for payment was never reached.[13]

When Heinze's landlord discovered the ploy, he promptly sued. Heinze hired the best lawyer he could find—and won. It was Fritz's first lawsuit and it would not be his last.

With diabolical brilliance, Fritz married his knowledge of Butte's geology with a pernicious mining law known as the Apex Rule. According to the Apex Rule, rights to underground mineral holdings were determined by the location where a particular vein came closest to the surface, or reached its apex. If a company owned the apex, it could follow the vein below ground to *wherever it led*—including ground beneath another owner's surface property. Using his knowledge of mine engineering, Heinze began to identify and purchase strategic apex properties—many of them next to Anaconda (later Amalgamated) holdings. Then he pirated their ore. Lawsuits sprang up like weeds.

Lawyers, it turned out, would be Heinze's most important mining tool. In Butte they called him a "courtroom miner." Heinze fielded an army of more than thirty-seven lawyers, some of whom were charged with the lucrative task of digging up new opportunities to file claims and/or sue. The highly fractured nature of Butte's mineral veins created a lawyer's paradise, and Fritz learned a lesson that litigators still apply today: An expert can be hired to say anything. Elaborate scale models, "some costing $25,000 apiece," were built to support testimony. In other instances, testimony was supported by "litigation drifts"—holes dug for the sole purpose of creating courtroom evidence.[14]

Heinze filed his most notorious apex claim in the spring of 1899, just as Butte was awakening to the threat posed by Standard Oil's purchase of Anaconda. Amid the convoluted patchwork quilt of mining claims on Butte Hill, Fritz and his lawyers somehow found two tiny parcels of unclaimed land. Added together, the two parcels equaled less than one *one-hundredth* of an acre—"the equivalent of a small room." At this minuscule point of land, claimed Heinze, lay the apex of the great Anaconda vein. Revealing his penchant for hilarious irony, Heinze named his new mine the "Copper Trust." It was, sputtered the *Anaconda Standard,* "an astounding piece of audacity." So audacious that Butte's courts, in a rarity, actually failed to support Heinze to the hilt.[15]

Most of Heinze's claims, however, found a sympathetic hearing. This did not occur by accident. Whether or not Heinze actually bribed judges is not clear, but it is certain that he worked tirelessly to ensure the election of his allies to key Montana courts. Perhaps his greatest asset was a colorful Butte judge named William Clancy. Clancy had a cartoonish appearance dominated by a tangled, flowing gray beard. According to one story, a man once approached Clancy and said, "Judge, I'll bet you a dollar that I can tell you what you had for breakfast this morning."

"You're on," said Clancy.

"Ham and eggs," said the man triumphantly, lifting Clancy's beard to point out the physical evidence.

"You lose," said Clancy. "That was yesterday's breakfast."[16]

Clancy was just as notorious for his outlandish courthouse behavior. He was often manifestly disinterested in legal proceedings, a fact he sometimes demonstrated by sleeping. When awake, he could be caustic and withering toward courtroom participants—especially those affiliated with Amalgamated. Once, irritated with a particularly loquacious witness, Clancy erupted: "It seems to me this witness is troubled with constipation of thought and diarrhea of words."[17]

Whatever anyone else might think of Judge Clancy, for Fritz Heinze he was a sure bet. It was a truism with which Standard Oil/Amalgamated would soon become painfully familiar.

F ritz Heinze's first major engagement against Standard Oil came in the context of the election of 1900. Heinze's goal in this election flowed directly from his signature method of extraction. As a "court-room miner," Heinze depended upon friendly courts; he set about to ensure that his allies occupied key benches.[18]

Like any savvy political strategist, Fritz recognized that there is no asset like a good enemy. In this regard, at least, the Standard Oil purchase of Anaconda was like a dream come true. Nationally, "progressivism" was taking off as a powerful political movement—a reaction, in part, to the rise of abusive monopolies. Montanans, on the heels of a decade in which populist sentiment had run strong, resented and even feared the idea of control by the most notorious of eastern trusts.[19]

Whether Heinze was a true believer or a clever opportunist is subject to debate, but in either event, he had found a potent rallying cry: "My fight against Standard Oil is your fight," Fritz told audiences across the state. "In this glorious battle to save the State from the minions of the Rockefellers and the piracy of Standard Oil you and I are partners."[20]

In his enmity to Standard Oil, Heinze found a powerful ally—or perhaps more accurately, a marriage of convenience. Copper King William A. Clark, a man incapable of shame, strode unabashedly back onto the stage. For Clark, noisy opposition to Standard Oil/Amalgamated would provide a satisfying slap at his bitter rival Marcus Daly. But Clark sought an even greater revenge: a new path to the prize he had long coveted and long been denied—a seat in the United States Senate. While Fritz Heinze saw the election of 1900 as a way to win friendly courts, William Clark envisioned a way to win a friendly state legislature—one that would elect him (legitimately, this time) to the U.S. Senate.

Between Heinze's political savvy and Clark's bottomless supply of cash, the two formed a potent team. Together they turned the election into a dramatic political circus. In a way it looked thoroughly modern— hired talent pumped into the state in a seamless melding of politics and entertainment: political cartoonists, famous actresses, fireworks, free champagne, and even vaudeville performers, slamming Standard Oil to a catchy ragtime beat.[21]

Standard Oil, hardly a naif, did not take the Heinze-Clark challenge sitting down. It began its own free-spending campaign, including a $1.5 million war chest to buy Montana newspapers.[22] Though the newspapers would be important later, they were not enough to swing the election of 1900. In the end, Heinze elected and/or re-elected a slate of friendly judges, and a new Montana legislature— barely a year beyond the bribery scandal of 1899—sent William A. Clark back to the United States Senate.

Within weeks, though, the ground beneath Fritz Heinze began to shift. Standard Oil employed a tactic that would serve it (and its successors) to great effect: Divide and conquer. Senator William A. Clark provided an easy target, devoid of scruples and vulnerable to his own recent past. Standard Oil threatened to dedicate its boundless resources to a revived U.S. Senate challenge of Clark's credentials— unless Clark broke with Heinze and dropped his opposition to Amalgamated. For Clark, it was an easy call. He shed himself of Fritz Heinze with the ease that most people change a shirt. Heinze now stood alone against the preeminent economic power of his day.[23]

In the immediate aftermath of the 1900 election, another development seemed to sharpen the battle lines. On November 12, Marcus Daly died. His death, wrote historian Michael Malone, seemed "to symbolize the passing of the great captains of frontier industry and to herald the emergence of the giant and impersonal new supercorporations."[24]

From 1900 to 1906, the second War of the Copper Kings raged through Butte and ultimately through all of Montana. It was a war with dozens of battles, but two in particular are worth noting. One brought Heinze and Amalgamated into literal warfare—on a battlefield hundreds of feet below ground. The other represented, in essence, a Standard Oil coup d'état of the once-sovereign state Montana.

The armed conflict broke out as a by-product of one of Fritz Heinze's infamous apex claims. Just before the turn of the century, Heinze had filed a series of claims based on his ownership of a mine called the Rarus—which bordered Amalgamated property rich in copper. Pending resolution of the case, both Heinze's Montana Ore Purchasing Company and Amalgamated were enjoined from mining in the contested zone.[25]

In 1903—with the case still pending and the injunction still in place—Heinze executed one of his most outrageous legal maneuvers. He simply transferred ownership of the Rarus mine from his Montana Ore Purchasing Company to one of his other corporations, the Johnstown Mining Company. Then, with the disingenuous reasoning that the Johnstown was not subject to the court's injunction, he secretly began to mine the contested ground.[26]

Over a period of many weeks, Heinze flooded miners into the Amalgamated holdings, pulling out hundreds of thousands of dollars' worth of copper. Amalgamated finally caught wind of the scheme when its miners, working in a nearby property, began to hear blasting—clearly coming from the "forbidden zone." Their suspicions finally raised, Amalgamated sent spies on an underground mission to determine the scale of the pirating. They were shocked at the wholesale plunder.[27]

Back on the surface, Amalgamated asked a federal judge to mandate an inspection of the property. But before the inspection took
place, Heinze covered his tracks by dynamiting the cavities and filling
them with waste rock. More injunctions were issued, which Heinze
now blatantly ignored, continuing his extractions from the forbidden
ground.[28]

Below ground, meanwhile, the struggle degenerated into open warfare. Fistfights broke out when groups of Heinze miners encountered
their counterparts from Amalgamated. These quickly escalated, and
soon miners were fighting one another with more dangerous weapons.
Dueling miners burned "stink-producing material"—including rub-
bish, old shoes, and rubber—to fill the enemy's workings with smoke.
They blew a powdered form of lime through air hoses to pollute the
tunnels with caustic dust. They diverted water to cause flooding and
used dynamite to cave in tunnels and block access points. The miners
even tossed homemade hand grenades—short pieces of dynamite
crimped inside tomato cans.[29] Remarkably, amid weeks of running skirmishes, only two men were killed.[30]

Poised on the brink of large-scale loss of life, both sides seemed to
catch themselves and called a truce. After stealing more than
$500,000 worth of copper, Heinze would ultimately pay a fine of
$24,000—less than 5 percent of his haul.[31]

While dramatic, the Rarus apex case was just the tip of Heinze's gigantic litigation iceberg. Some lawsuits Heinze won outright. With others he secured injunctions, idling Amalgamated properties while
endless proceedings plodded through the courts. Amalgamated was
spending millions in legal fees, and even more significant, Heinze's suits
had tied up Amalgamated holdings valued at $70 to $100 million.[32]

Building on his success in the election of 1900 and his devastating
legal attacks, Heinze also mounted a relentless public relations barrage.
He continued to cast himself as the local David against a heartless,

foreign Goliath, and the people of Butte rallied to his side. F. Augustus Heinze was in many ways like a Butte version of Robert E. Lee—a beloved field general reeling off a remarkable string of battlefield successes against a much larger foe. Like Lee, however, Heinze faced an opponent with patience, resolve, and vastly superior resources. And for Heinze, his Gettysburg was about to unfold.

Apex litigation may have been Fritz Heinze's trademark, but it was far from the only weapon in his legal arsenal. Heinze was also giving Standard Oil/Amalgamated fits through the technique of minority shareholder suits. In a minority shareholder suit, Heinze allies purchased a few shares of companies being absorbed into Amalgamated. Then they devised claims as to how Amalgamated was violating their minority shareholder rights and brought suit—preferably in a court presided over by a close Heinze ally.

In the most far-reaching claim, Heinze's minority shareholder allies argued that the absorption of component companies into Amalgamated had been undertaken without their consent—and therefore should be quashed. As a holding company, Amalgamated was nothing without the companies it had absorbed. If Fritz and his allies won, Amalgamated would effectively be dissolved. Heinze, of course, had chosen his forum carefully. The suits were filed in the court of an old friend—the gray-bearded Judge Clancy.[33]

On October 22, 1903, Judge Clancy ruled in favor of Heinze, prohibiting Amalgamated from absorbing component companies (without the permission of *all* shareholders) and also prohibiting the Company from paying dividends derived from the profits of the acquired companies.[34] If Clancy's ruling stood, Amalgamated was dead—along with Standard Oil's broader play to build a copper trust.[35]

With the corporate stakes now life or death, Standard Oil came roaring back. *Montana wants Amalgamated shut down? Show Montana what that means.* Within hours of Clancy's decision, Standard ordered a total shutdown of all Amalgamated operations in the state. Amalgamated mines in Butte were immediately closed. The great smelter fires were extinguished in the towns of Anaconda and Great Falls. Across Montana's forests, lumber camps were emptied. So too were sawmills, coal mines, railroads, retail stores, and dozens of other related industries. Overnight, three-fourths of Montana's wage earners found themselves abruptly thrown out of work.[36]

Having shown Montanans the price of defiance, Standard Oil issued its demand to Heinze: Sell the offending minority shares and drop the suits. Heinze, playing for time, promised to answer the next day from the courthouse steps.

What followed was one of the great pieces of political theater. When Fritz Heinze arrived at the appointed hour to give his response, an audience of *ten thousand* people stood waiting outside the courthouse. It was an angry crowd. Fritz, he knew, had tapped too deeply from his reservoir of goodwill. His injudicious use of lawsuits had triggered a terrible backlash, and now thousands of workers—including the teeming mob that stood before him—faced a frigid Montana winter with no jobs.

Heinze began to speak, his deep voice booming so that his words could be heard up and down the street. He deplored the control that Amalgamated had gained over the affairs of the state. The emerging trust, he argued, was the "greatest menace that any community could possibly have within its boundaries." He reminded them that Amalgamated was really Standard Oil, and that Standard Oil had "trampled every law, human and divine."

Most of all, though, he cast *his* fight as *their* fight.

If they crush me today, they will crush you tomorrow. They will cut your wages and raise the tariff in the company stores on every

bite you eat and every rag you wear. They will force you to dwell in
Standard Oil houses while you live, and they will bury you in
Standard Oil coffins when you die.

It was brilliant demagoguery, all the more effective because—at some level—it was true. So great was Heinze's oratory that, for a few days at least, he managed to turn the tide back in his favor.

Soon, though, Standard Oil issued a new demand—so audacious that it made the old William Clark bribery scandal look like child's play. This time the great trust ignored Heinze and went straight to the governor of Montana, Joseph K. Toole. In exchange for reopening Amalgamated's operations, Standard demanded that the governor convene a special session of the Montana State Legislature. Once convened, the legislature was directed to pass a new law that would allow offensive judges to be disqualified in civil suits. They called it the Clancy Law.[37]

Governor Toole, to his credit, initially resisted, rightfully horrified at the precedent of acceding to such raw corporate blackmail. Ultimately, though, with winter looming and most of his state out of work, the political pressure was too great. On December 1, 1903, the Montana legislature convened in special session. By December 10, Standard Oil had its new law. Fritz Heinze had lost his most potent weapon—the courts. Heinze would mount a few skirmishes in the ensuing months, but the end was now written.[38]

In 1906, F. Augustus Heinze sold his Butte holdings to Amalgamated for $12 million. In 1910, William A. Clark followed suit, selling most of his property in Montana. Through Amalgamated—in 1915 the name reverted back to the old "Anaconda"—Standard Oil had consolidated virtually all the holdings of the three former Copper

Kings. The new corporation was a leviathan of remarkable scale and power. "Regardless of its shifting corporate entity," said one Montanan who watched the changes, "it was always referred to . . . as 'the Company,' a simple yet awe-inspiring term."[39]

Marcus Daly, of course, died before seeing the evolution of the company he built. Heinze moved to New York City, where he lived in a double suite at the Waldorf, married a beautiful actress, squandered most of his fortune in a banking venture, then died at age forty-five from cirrhosis of the liver. William Clark, after a single term in the U.S. Senate, would also spend most of his time in Manhattan (when not in Europe), collecting art and residing in his 121-room Fifth Avenue mansion.

Back in Butte, meanwhile, the miners and their families would live with the consequences.

One of the men who would soon feel the consequences of Standard Oil's Butte takeover was Burton "B.K." Wheeler. By the time of the North Butte disaster, Wheeler held the powerful position of federal district attorney for Montana. He had arrived in Butte, though, as a freshly minted graduate of the University of Michigan Law School, in the waning days of Fritz Heinze's epic battle with Standard.[40]

B.K. Wheeler, destined to be one of the most important men of his generation, settled in Butte as the result of a losing poker hand. Wheeler was born in Hampton, Massachusetts, in 1882, the son of a cobbler. He worked his way through the University of Michigan Law School, clerking in the dean's office during the school year and selling cookbooks door-to-door in the summer. Upon Wheeler's graduation, the dean offered to help place him in "one of the big New York law firms."

B.K. Wheeler declined. "[R]eturning East seemed stultifying," wrote Wheeler in his autobiography. "I was anxious to go anywhere that was wide open with opportunity . . . ever since I was a child I had dreamed of going West." With his life savings of $500, Wheeler set out to find a job, hopping from town to town for three months before arriving in Butte in the fall of 1905—just as Standard Oil was fixing its grip on the city.

Wheeler spent a week interviewing with "every successful lawyer in the city." The results of his effort: "exactly one offer." And not one he found attractive. Wheeler packed the bag containing his worldly possessions and headed for the train station. "I decided to try Spokane."

On his way to the station, two well-dressed men, presenting themselves as fellow travelers, invited the young man to share a drink. Wheeler accepted the invitation, and as they sat down at a bar the men proposed "a friendly game of cards." The young lawyer suggested "pitch."

"Oh no," said one of his companions, "let's play poker."

In a matter of minutes, Wheeler was betting the remnants of his life savings on a pair of jacks. When one of his new friends turned up three treys, the game was over. "I sat there dumbfounded," said Wheeler. He learned later, of course, that the game had been rigged, but the result would stand. The train for Spokane came and went, and Wheeler accepted the job he had rejected the day before.

Eight

·

"THEN WE MET DUGGAN"

Then we met Duggan, a nipper, who knows every foot of the ground.
—JOSIAH JAMES, QUOTED IN THE *ANACONDA STANDARD*, JUNE 11, 1917

Miner John Wirta was at work on the 2,600 level of the Speculator in the early morning hours of June 9 when "an Austrian" ran through his crosscut, yelling in broken English, "Come on! Come on! Pretty danger!" At first Wirta and his crewmates thought the man was crazy. "But we feared there might have been a cave-in or fall of ground above, so we went up to the twenty-fourth level." At the 2,400, according to Wirta, "We smelled smoke and gas for the first time and heard that there was a big fire in the shaft."[1]

Though their situation was obviously serious, Wirta and the others with him could take one bit of comfort. The 2,400 level of the Speculator connected directly to the High Ore mine. One of the men knew the passage and "thought we could get out that way." They took off through the connecting drift.[2]

The miners had progressed only a few hundred feet when they encountered a concrete wall with a rusted iron door, sealing the drift closed. They tried the handle, but the rust froze it solid. Precious minutes ticked away as they pounded at the handle, the smoke and gas pursuing them through the drift. There must have been collective elation when finally the door gave way, swinging back on its corroded hinges toward the Speculator.[3]

But as the men turned their lanterns to illuminate the opening, they did not find a passageway to safety. Directly behind the door stood another concrete wall, blocking their path from top to bottom and from side to side. Desperately they searched for tools—something to batter down the wall. Finding nothing but train tracks, the miners pulled up the rails and began to pound. The wall, though, was too thick: four inches of concrete framed by a one-inch timber brattice.[4]

The experience of Wirta and his companions was not unique. In numerous North Butte connections to adjoining Anaconda properties, miners fleeing the fire encountered bulkheads—solid walls obstructing their escape. Many of the bulkheads had apparently been built after the May 1917 fire in the Anaconda-owned Modoc mine. Ironically, the intent of the walls had been to keep the Modoc smoke and fumes out of the Speculator.[5] In this regard, the structures were perfectly proper. What was *improper*—indeed what was *illegal* under Montana State law—was for walls to be built *without doors*.[6] Some dead miners were found "piled up against bulkheads of solid cement," remembered B.K. Wheeler, the federal district attorney, "although state law required that all bulkheads in the mines must have an iron door which can be opened."[7] The rationale for requiring doors, of course, was to prevent the precise scenario that was playing out in the Speculator—the entrapment of men in the event of an emergency.

The contemporaneous accounts do not record the terror and panic that Wirta and the others must have felt as they battled this new obstacle, but bulkheads built without doors would emerge as one of the most bitter issues in the aftermath of the fire. There were widespread, graphic rumors that rescue crews found dead miners with "their fingers worn to the knuckles in an attempt to reach safety."[8] Wirta's group would manage to avoid at least this fate. Grasping the futility of pounding at a concrete wall without proper tools, they ran back into the smoky Speculator.

"We first tried to go up to the twenty-second level," reported Wirta, "but the gas was so bad that we were forced to stop." Back down at the 2,400, Wirta gathered twenty-five sticks of dynamite, determined to return to the High Ore bulkhead and blast his way through. But it was too late. By then strong gas had filled the connecting drift, preventing all access. "We all thought that we were facing death," said Wirta. "This was the first I heard of the nipper."[9]

Wirta would emerge as one of the key chroniclers of the North Butte disaster for an important reason in addition to his good memory. He was one of the few miners who carried a watch.

M ore men began to converge in the main tunnel at the 2,400 level of the Speculator. Like John Wirta, most had already made other attempts to escape and, like Wirta, most believed that their options had dwindled to nil. Two of the men joining this growing group were Albert Cobb and his partner, Henry Fowler. Barely an hour earlier, Cobb and Fowler had been working in the 2,400 Station of Granite Mountain when Ernest Sullau came scrambling up from the shaft, burning cable at his heels, desperately demanding a bucket of water.

When it became clear that the Granite Mountain fire would not be doused by buckets, Cobb and Fowler fled in the direction of the Speculator, warning others as they moved through the crosscut. Before long, though, they encountered a group of miners fleeing in the opposite direction who told them "the station was bulkheaded." Next

Cobb and Fowler tried to climb up a level but found that path blocked by gas as well. "We were like a bunch of fools and did not know where to go," remembered Cobb. "We met Duggan after we came back from the Speculator shaft."[10]

M anus Duggan was born on May 30, 1887, in Coatesville, Pennsylvania.[11] He was pure-blooded Irish, the son of two first-generation immigrants named Mary and John Duggan. The closest that Manus came to a birthright was his auburn hair, green eyes, and a tradition of mining—his father made his living in the Coatesville coal mines. At the time of the fire, Manus's father was dead. His mother, though, was actually on a train bound for Butte. Manus had saved for months to raise the money to bring her west.[12]

As a boy, Manus had managed to obtain an elementary education. At age twelve, though, he joined his father in the coal mines, "picking slate."[13] The precise year that Duggan arrived in Butte is unknown, but it is believed that he headed west during one of Coatesville's periodic downturns. He may have arrived in Butte in 1906 at the age of twenty-one. It is known that in that year Duggan took a room at the Brogan Boarding House, an establishment named for its proprietor, Mary Brogan.

Mary Brogan's daughter Madge was a preteen when Manus moved in, but he caught her eye from the beginning. "He was the finest looking man who ever walked the earth," she would say in her later life. "I was crazy about him from the time I was eleven years old."[14]

Madge was eighteen and Manus was twenty-seven when they married, on April 7, 1915, at the Sacred Heart Church. Forgoing any sort of honeymoon, the newlyweds put Manus's modest savings toward the construction of a small house—which Manus built himself.[15] It was a piecemeal process, with construction progressing in tandem with Manus's wages. Madge would remember the greatest gift that Manus ever gave her—a porcelain commode. The indoor toilet took the place of an outdoor privy and eliminated frosty nighttime treks during Butte's frigid winters.

Duggan worked for the North Butte Mining Company as a "nipper." The primary responsibility of nippers was to sharpen tools, a constant need, as the miners drilled into solid rock. In the course of their workday, nippers moved throughout the mine, gathering up tools, shuttling them to the surface, and returning with freshly honed equipment. In their daily travels, the nippers developed a familiarity with every crosscut, manway, and drift.

A miner named Josiah James was the first to encounter Manus Duggan after the start of the fire. James was working with his partner, a young Italian, when they learned of the burning shaft. The two partners, after fleeing the burning Granite Mountain shaft, decided to split up in order to warn others. The Italian miner would later be found dead. James had better luck—the first man he met was Duggan.[16]

Duggan offered to help spread the alarm as they searched desperately for a way out. Manus and James ran through the smoky labyrinth, gathering up the men—and boys—they found along the way. They covered an enormous amount of ground in a period of less than an hour: down twenty stories from the 2,400 to the 2,600; back up to the 2,400; then up to the 2,200; then down again to the 2,400. Seventeen-year-old Willy Lucas was at work on the 2,400 when he received the nipper's warning, falling quickly in tow.[17] At the 2,600, Duggan found an Austrian immigrant named Godfrey Galia. "I was working on the 2,600 level when the nipper ran in and told us there was a fire in the shaft."[18] A man named Charles Negretto remembered seeing Duggan "waving a light" as he and others "were running as though wild" through the drifts.[19]

Sometime around 1:00 A.M. on the morning of June 9—an hour after the start of the fire—Manus Duggan and twenty-eight other men converged on the 2,400 level of the Speculator. Collectively, the men had tried escape routes through the High Ore, the Rainbow, and the Speculator. They had tried climbing up and they had tried climbing down. But everywhere they went they found smoke and gas—and thicker by the minute. There was no way out. "The gas was too strong and the crowd stopped for a consultation," remembered Josiah James.[20]

Dread fell over the miners. A few men debated arming themselves with dynamite and making a final, frantic dash—blasting through any obstacle they encountered with the blind hope of eventually finding passage to the surface.[21]

"Boys, you can suit yourself," said Duggan. "You can try to get up the thirty-seven raise through that gas or go over to the Speculator, but I am going to stay here and bulkhead myself in." He looked around at the men surrounding him. "I advise you to stay with me."[22]

It was a stark proposal—entombing themselves, half a mile below ground, in a burning mine, choked with poisonous fumes—but there was no other way. Duggan knew of a "blind drift" nearby, a dead-end tunnel where he believed the air was still pure. Directing the men to gather building materials as they fled, Manus and the twenty-eight others began to run.[23]

The drifts were strewn with potentially useful items, and the miners grabbed everything they could put their hands on. "[W]e each took a lagging and what timber we could get hold of," remembered Cobb. The men found a large canvas pipe used for ventilation, and Duggan yelled for them to tear it down. Others grabbed saws, shovels, and nails. Cobb and Fowler remembered having seen a barrel of water, but when they went to retrieve it they discovered it had been spilled. Wirta and another man did find a keg with two gallons of drinking water. Someone pulled in a "submarine," the miners' term for a small tank of water on wheels, partially full of thick copper water.[24]

The blind drift where Duggan led them extended about 500 feet into the rock, with its mouth about a thousand feet from the 2,400 Station of the Speculator. The location that Duggan chose for building the bulkhead was about halfway into the drift, leaving approximately 250 feet of tunnel space behind it.[25] When they reached this spot—a little after 1:00 A.M.—the air there was good. "[T]he gas had not got in there yet," remembered the boy, Willy Lucas.[26]

For a moment, the situation seemed somewhat less grim. The men were winded from the sprint, and one of the miners suggested "taking five." Duggan flatly rejected the notion of a break, directing the men instead to "make all speed."[27]

Manus supervised construction, with four men at a time working in the narrow space. The miners would later agree that their work product was "the most strangely constructed bulkhead in the history of the local district." Looks, though, mattered little, and Duggan's design was proof that necessity is the mother of invention. He used the timbers and lagging to build a two-sided wooden frame with about eight inches of open space in the middle. It looked like a crude version of the forms built before concrete walls are poured. Duggan, of course, had no concrete. Fowler described their substitute: "We used our sweaters and jumpers and other clothing to stop up the chinks, and threw dirt into the cracks to keep the gas and smoke out."[28]

When the sweaters and jumpers ran out, the men added their shirts, socks, shoes, hats, even their underwear. "In making that bulkhead the men gave up most of their clothing," said Leonard Still, another of the men who was there. "Some gave up everything, and did not have a stitch left."[29] They left a hole on the side of the bulkhead, plugged with a shirt, to allow them to check the outside air.

It took Duggan and his men less than an hour to complete the bulkhead. Before they were finished, the men began to smell gas. And by the time they were done—around 2:00 A.M.—smoke was filling the drift in front of their wall. Cobb was the last man to climb behind the bulkhead before they sealed themselves in. With a pencil he scratched three words on one of the outward facing boards: "Men in here."[30]

The men were scarcely behind the bulkhead before Duggan directed them to build a second wall, ten feet away.[31] The second bulkhead was less elaborate—probably constructed of canvas. It had some type of a flap that allowed passage from the main chamber into the space in between the two bulkheads.

Duggan, from the moment he stated his intention to build a bulkhead, commanded the group. In their later accounts of the disaster, the men with Duggan would make frequent reference to his "orders" and "commands" as if he were the leader of a military unit. There was not the slightest hint of resentment, since at least he had a plan. Indeed, Duggan's authority rested purely on merit—earned through his superior knowledge of the mines and his sturdy confidence in the course they should take. As a nipper, Duggan occupied a low rung in the informal ranks of the mines, yet his quiet poise inspired faith, and no one questioned his command. At least not yet.

One of Duggan's first orders, once the miners finished construction of the second bulkhead, was the establishment of a formal rotation among the men. Willy Lucas described it: "Two men were in the space between the two bulkheads all the time with a box of dirt to plug the holes so the gas could not get through. They were not the same two men, but would stay for an allotted time before two others would take their places."[32] The two men on duty would also bang against the railroad tracks and a pipe on the wall, hoping to attract the attention of any rescuers in the area. At the end of their rotation, the two men moved from the space between the two bulkheads to the *back end* of the blind drift—in effect, the end of the line. The rotation would ultimately prove vital in ways that Duggan himself did not likely foresee.[33]

With the bulkheads built and the rotation established, the twenty-nine men settled in. They occupied a drift six feet tall, six feet wide, and 250 feet long. Along the length of their tunnel ran a railroad track, and most of the men lay or sat along the rails. The ground beneath the men was dry. The temperature behind the bulkhead was eighty-four degrees and the humidity was 88 percent—quite warm but not stifling. (The men, in any event, had been liberated of most or all of their clothing.) Though deep in the earth, the bulkhead was well lit. Many of the men had their carbide lanterns, blazing away. John Wirta, the man with the watch, was a rare miner carrying an electric flashlight—a possession that would later play an important role.[34]

In the early hours behind the bulkhead, the mood was surprisingly buoyant. The miners, in typical fashion, cracked jokes. Perhaps mocking the earlier request, one of the men suggested that they "take five and lunch."[35] Some of the men sang songs or hummed, and a few even played cards. Hunger was not a problem, because most of the men had eaten just before the fire. Nor was water yet a concern.[36]

It was a quiet interlude in the early hours of a long ordeal. Within twenty-four hours, Duggan and his fellow miners would find themselves pushed to the brink of death.

Nine

•

"HEY JACK, WHAT THE HELL"

Hey Jack, what the Hell. We're here today and gone tomorrow!
—*COPPER CAMP*[1]

The men behind Manus Duggan's bulkhead, like the broader community of miners, arrived in Butte via diverse paths. For all of them, though, the destination was the same—a job.

Many of the men carried mining in their blood—Duggan hailed from the coal fields of Pennsylvania. Others were sons of miners from the once-dominant copper fields of Michigan, or the silver mines of Nevada's Comstock Lode. In 1917, many Butte miners could trace their lineage to grandfathers who had chased the dream of gold in California or Colorado. Europe too sent her sons to Butte, none more famously than the "Cousin Jack" Cornish and the Irishmen of County Cork. These second- and third-generation miners were the aristocrats of the mining world—a world that was transforming around them.

By 1917, the mining industry was in the midst of dramatic change. Consolidation, competition, immigration, technology—all had combined to shake the foundations of the Butte miners' world.

Since by 1906, Standard Oil had consolidated most of Butte's productive mining properties into the single megacompany, Anaconda, gone were the local owners who viewed their workers with a sort of benevolent paternalism. Gone were the local rivalries that forced the owners to compete for the loyalty of their men. Gone too—as events would soon underscore—was the powerful unionism that had helped to ensure significant reward in exchange for the enormous risk inherent in the miners' daily work.

Though Anaconda was omnipotent locally, Standard Oil had failed in its goal of creating a broader copper trust. Competition from other states—especially Arizona and Utah—was fierce. By 1917, in fact, Montana had lost its long-held dominance of the copper industry, with Arizona assuming the position of the nation's top copper-producing state.[2] Competition was also beginning to emerge from overseas—especially Chile. In 1915, the Guggenheim family began the operation of a well-funded, massive new copper enterprise at Chuquicamata. Indeed, exploration would ultimately reveal that the Chuquicamata reserves were greater even than Butte's "richest hill on earth."[3]

One reason for the growing success of Arizona, Utah, and Chilean mines was their use of new mining techniques and technology. In the 1880s and 1890s, the Butte Copper Kings had driven their Michigan competitors into the ground through technological prowess. Now it was Butte that was scrambling to keep pace. By 1917, the most modern mines were using "open pit" techniques, abandoning the strategy of sinking narrow shafts in a focused search for the elusive vein. In open pit mining, they simply dug up everything. In Chile, the Guggenheims had purchased the gigantic steam shovels that had formerly been used to dig the Panama Canal.[4]

Butte could not apply open pit techniques because—uniquely—a city stood atop the underground copper. To compete against foreign resources and open pits, Anaconda put other new technologies to work—most notably mechanical drills. The skilled older miners found that this new technology required more brawn than technique, a fact that made it increasingly easy for Anaconda to put unskilled workers into the mines.

Changes in immigration trends, meanwhile, provided a steady stream of cheap labor. Butte had always been an immigrant community—especially since the beginning of large-scale industrialized mining in the silver era. But by 1917, the ethnic makeup of Butte's immigrants had undergone dramatic change. The Irish, though still dominant, watched the arrival of wave after wave of new immigrants, including Finns, Italians, Russians, and Eastern Europeans. In this new Butte, "Dublin Gulch" was now a neighborhood of "bohunks"—Croatians, Serbs, Montenegrins, and Slovenes.[5]

In a classic pattern, the established resented the newcomers. The recent arrivals were willing to work for lower wages—a clear threat to job security. Following a 1915 accident a handbill was circulated reading "Be Careful. Don't Get Hurt. There Are Ten Men Waiting For Your Job."[6]

Unskilled and unable to speak English, newly minted miners were a safety nightmare, and in the unforgiving environment of the mines, novices put not only their own lives at risk, but also the lives of those around them. In contrast to German or Cornish mines—where new miners were inducted via a lengthy program of apprenticeship—Butte mines had no training program for new miners. In Butte, new men were simply handed a shovel and directed to muck. There are numerous anecdotes of miners who died on their first day on the job.[7]

World War I caused both demand and price for copper to spike, intensifying the frenzy for production. Anaconda hired an even higher number of inexperienced miners, and the drive to meet production quotas increased. Even without the North Butte disaster, the war years were among the Butte mines' most deadly.[8]

The men with Manus Duggan reflected Butte's ethnic evolution like a tiny microcosm of the broader community. Certainly the Irish and English surnames were still present: Duggan, Cobb, James, Shea, Stewart. But so too were the names of those more recent arrivals: Popovich, Wirta, Evcovich, Beyersich, Gosdenica.

When a Butte miner walked to work, he traversed one of the most god-awful ugly cities in America. Which is not to say there was no beauty in Butte. There was, but like the miner's job, it lay deep below the surface. For its citizens, the town's ugliness was practically a point of pride. Butte, they understood, was first and foremost a mining camp. It was organized like a mining camp—for short-term expediency, with the expectation of imminent abandonment. Somewhere along the line, its inhabitants decided to stay, building a cosmopolitan city on top of a mine.

The wealth of the copper industry brought Butte Victorian mansions, six-story department stores, plush theaters, and a hilltop college overlooking the town like a Butte version of a Greek temple. Yet interspersed among Butte's fine buildings was the cruder architecture of the mines. Dozens of gallows headframes marked the top of mine shafts dropping half a mile into the earth. Elegant buildings stood next to forty-foot piles of slag. Railroad tracks connected every mine to the nearby smelters, with trains shuttling through town around the clock.

Smoke from the smelters, meanwhile, was so thick that streetlamps were sometimes kept illuminated during the day. The smoke, loaded with arsenic, killed every tree—*every* tree, and every blade of grass. Except on St. Patrick's Day, Butte was devoid of green.[9] Copper King William A. Clark, painting a rosy picture of the poisonous air, once claimed infamously that "the ladies are very fond of this

smoky city, as it is sometimes called, because there is just enough arsenic there to give them a beautiful complexion." Clark also maintained that Butte doctors believed the smoke acted as a "disinfectant,
and destroys the microbes that constitute the germs of disease."[10]

Even the blind knew that Butte was ugly, because in addition to arsenic the smelter smoke was laden with sulfur, also known by its more
biblical name of brimstone. Butte, literally, smelled like hell. Nor was
smelting Butte's only noxious industry. The people of Butte had to
eat—and at the turn of the century there were five slaughterhouses
within a mile of downtown. The gory remnants of the trade were
hauled to giant lagoons on the edge of town. Open sewers, meanwhile, were another manifestation of lax city planning. One local entrepreneur investing in fly screens was said to have minted a small
fortune.[11]

A rriving at work, miners went first to the "dry"—a sort of locker
room where the men kept their work clothes. Putting on the stiff
and filthy clothing, remembered one former miner, was the "worst
part of the day."[12] The miners checked in at the timekeeper's office
before gathering beneath the headframe to await the cage. The clock,
for purposes of getting paid, did not start ticking until *after* the miners
had been lowered to their workstation.

The act of mining copper was at the same time simple and perilously complex. At almost every juncture in the process lay a significant possibility of death or injury. In 1915, sixteen North Butte miners
were killed before even setting foot on the hoist. They were waiting at
the collar while 600 pounds of dynamite was being loaded into a cage.
Somehow—some speculate a stray bullet—the powder ignited.[13]

Six or more men at a time rode in a cage, crammed into a space approximately two feet by three feet. If there was a prankster on board,

an experienced miner knew to keep one hand above his head as he squeezed in, because once the cage was full, it would be impossible to lift up an arm among the mass of men. Veterans would sometimes torture a greenhorn, his arms pinned to his side, by tickling his neck with a feather during the long ride down.

There was nothing amusing about the sometimes deadly consequences of riding in the cages. The hoists moved as rapidly as the elevators on modern skyscrapers—700 to 1,500 feet per minute. Unlike their modern counterparts, though, mine cages were open to the shaft. In 1911, eight Butte boys (all nippers, none over twenty) were killed as they rode a chippy cage to the surface. On board with them was a load of pipes. As they rocketed toward the surface, one end of a pipe somehow slipped through an opening in the cage. With the loose end clamoring and bouncing against the side of the shaft—the end inside the cage ground the boys to death. In another incident the same year, five miners were killed when the hoist engine malfunctioned, plunging the cage 1,500 feet to the bottom of the shaft. At the top end of the hoist, miners were sometimes killed when distracted operators failed to stop the cages at the collar—crashing them into the headframe.[14]

The basic act of blasting the rock from the mine walls involved drilling holes in a circular pattern, stuffing the cavities with sticks of dynamite, and igniting the charges. The loose ore was then mucked into small rail cars that were in turn rolled onto the hoist. The rail cars were lifted to the surface, where the ore was dumped directly into waiting trains, then shipped off for processing.

The use of dynamite, obviously, was rife with risk. "Tap 'er light" was a Butte colloquialism for "be careful." Its origin lay in the danger of tapping too hard when inserting dynamite into drill holes—the consequences of which could be loud and unhealthy. The flip-side risk was dynamite that *failed* to explode, and then had to be tenderly dug out of the drill hole by a miner with a knife.

Falling rock killed hundreds of miners. Miners always attempted to excavate in an upward direction, so that gravity would aid in bringing down the rock. The downside of this technique was that gigantic slabs of rock were sometimes left hanging precariously. The miners of Butte had a name for such dangling slabs—"Duggans." It had no connection to Manus but rather was the name of Butte's most prominent funeral parlor. To this day, miners around the world call a loose, hanging rock a Duggan—a mark of Butte's continuing influence on mining culture.[15]

Less obvious—but even more deadly—was the risk from drilling the holes into which the dynamite was placed. Before the turn of the century, a pair of miners drilled using a handheld bit and a sledge. One miner held the bit while the other pounded with the hammer. A small gauge bit was driven first, then gradually larger bits until the borehole was large enough to insert the powder. It required great skill—a slip would crush the hands of the man holding the bit. It also took a long time. County fairs of the day featured competitions rewarding the pair of miners who could drill the fastest hole. Hundreds of spectators would gather to watch the final rounds.

One of the technologies Anaconda applied in an effort to boost productivity was a new invention that the miners called "buzzies" or "widow makers." The buzzies were gigantic drills powered by compressed air that could bore a hole in a matter of seconds.[16] What made them deadly was the dust they generated—thick with fine particles of silica. When inhaled—which was impossible to avoid—the silica was devastating to lungs. Miners developed a deadly disease they called "miners' con" (for consumption). Doctors called it silicosis. In 1921, the Bureau of Mines published the results of an investigation of lung disease among Butte miners in the years 1916 to 1919. A staggering 43 percent had silicosis, and 6 percent had tuberculosis.[17] According to another study, 675 Butte miners died of respiratory disease in the years between 1907 and 1913.[18]

One analysis looked at miners' casualty rates *not including* lung disease. It concluded that if a miner spent ten years in the mines, he stood a one in three chance of being seriously injured and a one in eight chance of being killed.[19] In the decade *before* the North Butte disaster, mining accidents in Butte killed an average of one man a week.[20]

Even a miner's walk home was potentially hazardous. In his ride on the open-air hoist, he traveled from the depths of the mines—where men sometimes vomited from the heat—to the surface, which one miner described as "the coldest place I've ever been." In a five-minute winter ascent, it was not unusual for temperatures to plummet from more than 100 degrees below ground to zero or colder on the surface. Miners coming off the job were soaking wet with sweat, and the sudden exposure to Butte's frigid winter air led to a variety of ailments. This was particularly true before the advent of drys, which at least allowed the miners to shower and change clothes before walking home.[21]

Where a miner called "home" depended, in part, on his marriage status. Men tended not to marry until they had become somewhat established in their mining career. Manus Duggan followed a common pattern, marrying Madge a month before his twenty-eighth birthday—after working in Butte for more than a decade. One of the men trapped with Duggan on June 8, Wilfred La Montague, was scheduled to be married on June 11.

Many of Butte's married men owned their own homes, including Duggan. Duggan also built his own house—again typical. The homes of Butte miners were often constructed with materials pilfered from the Company, a petty crime for which miners felt no remorse. (Plumbers in modern-day Butte tell of houses in which no piece of pipe is longer than a lunch box.)

The majority of Butte's miners were single. Like Manus Duggan before his marriage, they generally lived in boardinghouses, some of them gigantic. The Florence, known as the "Big Ship," could sleep three hundred men. It was also common for store owners and private homeowners to rent out rooms. Room and board were generally offered together—including a daily lunch pail to carry into the mines. Sanitary conditions ranged from minimally adequate to wretched, and the larger dormitories were breeding houses for disease.[22]

When not sleeping, eating, or working, Butte miners of 1917 had access to a remarkable array of leisure activities. The risk and tension inherent in the men's daily work doubtless affected their taste for zestful entertainment. "Hey Jack, what the Hell," captured the prevailing attitude. "We're here today and gone tomorrow."[23] Naylor Wilson, who died in the North Butte disaster, was described as "typical of the care-free, fun-loving miner." The *Anaconda Standard* asked one of his friends for a comment about the dead man. Said the friend: "He probably found some enjoyment even during the fire."[24]

The stereotype of Butte as a "wide open" mining town was accurate and well earned. For every church in Butte there were six saloons, with names like Pay Day, Bucket of Blood, the Cesspool, and Open-All-Night. Because saloons were, indeed, open all night, most were built without locks. The Atlantic Bar, which claimed to be the largest in the world, was a block long and served up twelve thousand glasses of beer on a typical Saturday night.[25]

After a night of drinking, many miners headed for Venus Alley. As one bawdy house ballad put it:

First came the miners to work in the mine,
Then came the ladies who lived on the line.

In 1917, Butte's red-light district was said to employ a thousand women—one of the largest in the country. Despite periodic clampdowns, prostitution in Butte was openly acknowledged—and even tapped as a source of municipal revenue. Prostitutes paid a monthly licensing fee to the city.[26]

Beyond the mining camp stereotypes, Butte played host to dozens of more surprising cultural institutions. In 1917, Butte had fourteen theaters featuring fare that ranged from vaudeville to the new "photoplays" to the latest Broadway productions. During the week of the North Butte disaster, the newspapers ran ads picturing Al Jolson, complete with blackface. He and his 208-member cast were in Butte to perform the "original New York Winter Garden production" of *Robinson Crusoe*.[27] The show was accompanied by a live orchestra that played "The Star-Spangled Banner" at the finale.[28]

Though many miners were illiterate, many others were avid readers. In addition to three daily newspapers, Butte in 1917 had twelve libraries. The largest, the Butte Free Public Library, boasted 60,000 volumes and 2,000 new acquisitions each year. Ten thousand people held "book borrowers cards."[29]

Butte miners were members of all variety of formal "societies." One of the men with Manus Duggan was an Austrian named Godfrey Galia. Galia came to Butte via Roberts, Idaho, where he had participated actively in the local chapter of a service organization, the International Order of Odd Fellows. Butte, of course, also had its Elks and its Moose. Many societies were organized around ethnicity or religion and served as especially important social anchors for recent immigrants. There was the Scandinavian Brotherhood, the Servian Bocalian Brotherhood, the Ancient Order of Hibernians, and the Knights of Columbus. Dozens of other societies coalesced around commonalities ranging from Students of Nature to United Spanish War Veterans.[30]

In addition to participating in these more high-minded associations, Butte miners loved sports. In his free time, Duggan liked to ice-skate and, above all else, to play baseball. "He always had a baseball mitt on his hand," and in the evenings "walked out to Lake Avoca to play ball with the fellows."[31] Baseball had always been popular in Butte, and by 1917 there were scores of local teams. Football, wrestling, and Italian bocce were also popular, and even curling had a local club.

Butte miners held a particular soft spot for any sport that married athleticism with gambling, hence the wild popularity of horse racing and boxing. Crude boxing matches had been a Butte fixture since the earliest days of the camp. In one infamous 1889 fight, a carpenter and a miner agreed to settle a personal dispute in the ring. The men fought bare-knuckled for 105 rounds—two and a half hours. The following day, the carpenter died of his wounds.[32] Boxing became more refined over the following decades and soared in popularity during World War I. Butte's stature as a boxing town was sufficient to host title matches and reigning champions, including, in 1921, Jack Dempsey.[33]

Copper King Marcus Daly established horse racing as a popular Butte sport. His Bitterroot Valley horse farm helped provide world-class racing stock to Butte tracks—above all his beloved Tammany. Though the Copper King era represented the pinnacle, horse racing remained popular after Daly's death. A book about Butte published in 1943 bragged of the frequent appearance in Butte of jockey Red Pollard, "who was usually up on the immortal Sea Biscuit."[34]

The Butte newspapers devoted significant ink to sports coverage—local and national. One local celebrity was a well-known boxer/miner named Leo "the Butte Wildcat" Bens. Bens had been offered a job at the Speculator in the run-up to the fire. He turned it down because he "had work to do at home which would keep him busy for two or three weeks." There is no description of the household chore that kept him from the mine, but it may have saved his life. Bens would live to be ninety-seven.[35]

As for national sports stories, Butte papers in the summer of 1917 were fretting about a slumping Ty Cobb. "Is Ty Cobb slowing up?" asked one article. "Has he lost his batting eye?" Even as they worried (prematurely) about Cobb's decline, Butte miners also tracked a rising Red Sox pitcher by the name of George "Babe" Ruth.[36]

R eligion played an important and even central role in the lives of many miners. With the high numbers of Irish (and later French and Italians), Catholicism dominated the community. Dozens of other religions, though, were well represented. Butte in 1917 was home to forty-four churches (eight of them Catholic) and two synagogues.[37] Whatever their creed, though, most miners were also adherents to a less formalized doctrine—the doctrine of fate.

On the evening of Friday, June 8, an assistant foreman named Herbert Steinberg was working deep in the Speculator. As he labored in the mine that night, his father died. Word was sent to the mine and managed to reach Steinberg. The miner left his shift immediately, hurrying home to be with his family. A few hours later, the fire broke out, killing the men who had been working in his area. For all his remaining days, Steinberg would be left to ponder the eerie providence through which his life had been saved by the death of his father.[38]

Assistant Foreman Steinberg was not the only man to be saved by strange circumstance. A miner named Jack Fleck was hired at the Speculator in the week before the disaster, and his first shift on the job was to have been the night of Friday, June 8. On Thursday, however, "he had a little more money than he really needed," so he proceeded to "celebrate." His celebration was so active that he couldn't make it to work the next day, saved from the disaster by a hangover.[39] Newspaper accounts in the days following the fire document numerous stories of those who were saved by seemingly random circumstance: equipment failure, vacation, a trip to the surface for water.[40] The consequences of fate, however, cut both ways.

The men who perished in the disaster were led to death through their own long chains of mundane decisions. A miner named William Reichle was on his first day back in the mines after "an absence of several weeks from the city."[41] Twenty-one-year-old John Bixby was an engineering student at the University of California who had come to Butte in the days before the disaster, intending to spend the summer and "hoping to get some practical experience" before returning to college in the fall. The start of the fire on the night of June 8 coincided with Bixby's first-ever descent into the mine. His father would travel from Oakland to claim his body.[42]

Several men had the opportunity to contemplate their brushes with death in a particularly pointed way—having seen their names mistakenly included in the newspapers' lists of the dead. One young man, Frank Sangwin, faced the horrific task of identifying the "badly distorted" body of Harry Sangwin—his twin brother. Harry perished in the Speculator. Frank worked in the Diamond and escaped.[43]

All over Butte, the living struggled to understand the imponderable questions that arise with every disaster, struggled to make sense of the dividing line that separated those who survived from those who perished. And yet who could hope to explain it? It was not, after all, just the weak, or the stupid, or the evil that perished—but also the strong, the smart, and the good. What did it mean?

Some found comfort in the preachings of Butte's many churches and synagogues. Perhaps there was a higher purpose beyond the ability of mere mortals to understand. Perhaps the wave of sadness was countered, in some measure, by the acts of bravery and kindness in evidence across the city—above ground as well as below. Or perhaps, at least, a living God watched over the tragedy's aftermath—caring for the victims and weeping alongside their families and friends.

Some looked to superstition or astrology. One miner, John Lascovich, stopped working in the mines after a June 5 horoscope, three days before the disaster, warned of "a terrible explosion underground." He decided to go back to work after the fire but said he "would look at the horoscope each day it is printed," and that he would "have more faith in it in the future than he had before."[44] The *Butte Daily Post,* summing up an article that detailed the scientific explanation for death by carbon monoxide, wrote that June had always "been an unlucky month for Butte."[45]

Some found in the disaster a political cause—vowing to ensure that the deaths of the miners, however tragic, would not be in vain. Certainly the scores of dead were the direct catalyst of a strike that would soon force a crisis of national proportions.

Some, lacking any better explanation, simply fell back on "fate"— a vague notion that all of our actions are somehow preordained. Miners, in particular, seemed to find comfort in the idea that they walked along a path already carved out before them. A newspaper reporter asked one survivor of the disaster if he intended to go back in the mine. "Why, sure, that doesn't scare me," replied the miner. "We get used to these things. Some men are always being maimed or killed when they work underground. It might have been me. I may get it any time, but if I do I shall think it was my time to die."[46]

Ten

"IF THE WORST COMES"

*. . . if the worst comes I myself have no fears
but welcome death with open arms.*
—MANUS DUGGAN, JUNE 9, 1917

At 8:45 on the morning of Saturday, June 9, approximately seven hours after their self-interment, Manus Duggan scratched out a note by the light of the carbide lanterns burning indiscriminately inside the bulkhead.

> *Have been here since 12 o'clock Friday night. No gas coming through the bulkhead. Have plenty of water. All in good spirits.*[1]

Duggan would record his thoughts at two more junctures over the next twenty-nine hours, and his initially upbeat tone would decline dramatically in tandem with degenerating events.[2]

At some point, probably not long after Duggan wrote his first note, someone figured out that the carbide lanterns were robbing the men of their precious oxygen. Immediately they set restrictions, initially limiting the number of lanterns to four. Ironically, Manus had prohibited smoking in their early moments behind the bulkhead to protect the air quality, but he had overlooked their lamps. The limitations on the use of the lanterns helped to preserve the men's limited air supply, but it also forced the miners into ever-thicker murk.[3]

As the hours wore on, water too became an issue. The twenty-nine men had a total of only two gallons of drinking water—or less than nine ounces per man. While this amount could sustain survival for a limited amount of time, the men's discomfort grew with each hour. "We were up against it for water and that caused the most suffering," remembered Murty Shea.[4] Their thirst was made worse by the circumstances that led them to the bulkhead—an hour of running for their lives through the smoke-filled mine, followed by an hour of intense labor in building the bulkheads—all in conditions of heat and humidity.

Beyond the two gallons of drinking water, the men had only the submarine of copper water—the liquid used when operating Leyner drills to keep down the dust. Leonard Still remembered that "Duggan would not let any of us have any of the copper water, although all were thirsty."[5] Copper water was considered undrinkable. In some parts of the mine, in fact, copper water dripped from the walls. Men working in those areas had to wear protective clothing or else the seeping liquid raised sores on their skin.[6] Albert Cobb described the water's consistency as "thick like soup."[7]

The men entrusted the allocation of the water supply to Duggan, another indication of the esteem in which they held the nipper. All the men, though, seemed to rise to the occasion. When Duggan passed around the drinking water, "each man honorably took only his share and only a small portion."[8]

If there was a bright spot, it was the apparent soundness of their bulkheads. Duggan periodically checked the outside air, pulling the shirt plug on the side of the main bulkhead and putting his nose to the hole. Each time Manus checked—beginning an hour after they completed the first wall—the air outside was thick with the sweet smell of gas.[9]

Had the site for Duggan's bulkhead been selected with less knowledge or its construction completed with less skill, Duggan and all the other men would have died. Graphic illustration of this point can be found in the experience of a second group of North Butte miners.

Nineteen men at the 2,600 level found themselves in a situation similar to the miners with Duggan—no way out and no option but the construction of a protective bulkhead. For their building site, this second group selected a drift about 300 feet from the Speculator shaft. In choosing this location, though, the men made a fatal error. Unknown to them, a manway *behind* their bulkhead connected to the 2,800 level. The ground beneath their feet was also unstable, having recently been mined and then backfilled with loose material. Gas flowed into their chamber from both the open manway and the permeable floor. All nineteen men were found dead.[10]

Though the air behind the Duggan bulkhead was initially untainted, it was not unlimited. Use of the lanterns, at least, had finally been restricted. Still, with every inhalation, the twenty-nine men in the Duggan party depleted their collective supply of oxygen. And with every exhalation, the men breathed out carbon dioxide. Inevitably, with mathematically measurable certainty, the air supply became worse and worse.[11] As time wore on, "the gas from our breath was getting as bad as the gas from the fire."[12]

In addition to the diminishing oxygen and rising carbon dioxide, there was the additional matter of the sanitary conditions behind the bulkhead. Though none of the men discussed it directly, Willy Lucas gave a strong hint: "The air wasn't so very bad at first," said the boy. But after a while "it began to smell something awful."[13]

By Saturday evening, the oxygen content of the air behind the bulkhead had degenerated to such a degree that the men began to experience a noticeable difficulty in breathing. John Wirta, the man with the watch, remembered the time. "Saturday night about eight o'clock the men began to realize that they were dying a slow death."[14] They had been behind the bulkhead for eighteen hours.

For some of the men, it was too much. A man identified only as a "foreigner" named "Mike" apparently reached his breaking point. Wirta remembered him "jumping about in the drift, yelling like mad and stumbling over the rest of us, who were lying flat against the tracks to get every possible draft of fresh air."[15]

Accounts of the North Butte disaster, including newspaper stories and the official *Bureau of Mines Accident Report,* make frequent references to "foreigners" and "aliens" at work in the mines. They were vague terms, given the high rates of immigration and the diversity of countries represented in Butte. They also carried a clear pejorative tone, usually referring to non-English-speaking, recent immigrants from eastern and southern Europe—the latest waves to hit Butte.

There must have been an additional level of terror for these "foreigners" in the North Butte disaster. All of the miners, of course, were cut off from the surface. But the foreigners faced an additional degree of isolation—cut off from their fellow miners by their inability to speak English. For the foreigners, there could be no sharing in the collective knowledge about the nature and source of the fire; no contribution to the strategy for getting out; and no comfort from the reassurances of more experienced men.

Given what the men had already been through, there was no shortage of fodder for imagining apocalyptic scenes. Some of the men believed that "the whole mine and the whole town was on fire and that the best way was to bring death on suddenly and be over with it."[16] A few of the foreigners with Duggan were lucky enough to have a countryman by their side. Godfrey Galia would remember how he "spent most of the time lying there on the ground and holding tight to the hand of my friend, Martin Novak. Martin and I both came from the same part of Austria and each of us has a wife back there."[17]

The "Mike" who caused the disturbance was probably a Slav, Mike Evcovich. His effort to break out of the bulkhead may have been an attempt at suicide. Leonard Still gave this account: "One man named Mike tried to break down the bulkhead, and others joined him. We had to fight them off. I got a crack on the head while this was going on."[18]

Duggan took up a position to protect the bulkhead. Martin Novak called Duggan's action "the most heroic thing I ever saw underground." Novak described how Manus "stood with his back to the bulkhead and defied two men, who were apparently out of their heads."[19] The scuffle ended only when the two men finally "got exhausted and dropped."[20]

Duggan had weathered the first, half-crazed challenge against his strategy of staying behind the bulkhead. In a few hours, though, another challenge would arise—this time with the demonstrable support of the majority of the twenty-nine men.

From the time of Mike's outburst forward, conditions behind the bulkhead would worsen steadily. No problem was more elemental than the air quality—which continued to deteriorate. By midnight on June 9, according to Wirta's watch, "every man was gasping in long groans." They'd been entombed for twenty-two hours.

The mounting gloom was literal as well as figurative. It was probably around this time that the decision was made to reduce the number of carbide lanterns from four to one—one handheld lantern as the only source of light in a 250-foot-long cave. The men most likely kept this single lantern near the main bulkhead, so they could continue to monitor for leaks.

Manus had taken up a position near the bulkhead, from which he could guard both the water supply and the wall itself. Wirta remembered that Duggan at this time was "gasping for breath and muttering between breaths."[21] Another man described the "terrible pain that comes when your breath is shut off." The miners experimented with different head positions, kept their faces near the ground, used their hats as fans—anything to give them a few more breaths.[22]

Some men clearly believed that the end was near. "As I lay there with my head against a rail of the track I felt another man close to me," recalled Wirta. "He reached for my hand and pressed it."

"Good-bye, old man."

In recounting the incident, Wirta remembered a vividly sharp sentiment that he described as "the agony of impending death."[23]

There were more calls on Duggan to tear down the bulkhead. Duggan and Cobb made another check of the outside air. Instead of using the hole plugged with the shirt on the side of the bulkhead, they decided to drill a small hole near the top. "The air inside was stronger than the gas," said Cobb, "so the foul air went out for a while." Encouraged, they decided to try a hole at the bottom. "We figured there would be good air on the floor, but when we made a hole there the gas came in bad. We had to plug this up again and also the hole at the top." However dire the conditions behind the bulkhead, Duggan had clear reason to believe that the conditions outside were worse.

Yet the scattered demands to breach the bulkhead continued. To deflect these calls, Manus resorted to a combination of encouragement, stubborn refusal, and even subterfuge. "He would even lie about hearing sounds outside to keep the men back," said Cobb.[24]

At some point around this time, it appears that Duggan turned back to the short note he had begun to write in the early hours behind the bulkhead. The upbeat tone of the first installment was gone:

I realize that all oxygen has just been consumed. Everybody breathing heavily. If death comes it will be caused by all oxygen used from the air in this chamber.

By the time all the men were rounded together Friday night we were all caught in a trap. I suggested that we must build a bulkhead. The gas was everywhere. We built a bulkhead and then a second for safety. We could hear the rock falling and supposed it to be the rock in the 2,400 skip chute.

We have rapped on the air pipe continuously since 4 o'clock Saturday morning. No answer. Must be some fire. I realize the hard work ahead of the rescue men. Have not confided my fears to anyone, but have looked and looked for hope only, but if the worst comes I myself have no fears but welcome death with open arms, as it is the last act we all must pass through, and as it is but natural, it is God's will. We should have no objection.

He signed the note "Duggan."

Duggan was not the only man to seek comfort in a higher power, even as some of the trapped miners "blasphemed at the unhappy fate they were in." It was around this same time, recalled the men in sharp detail, that an "alien," in broken English, rose up on his knees and said a prayer aloud. In the simplest of terms, his prayer showed the mixture of fear, doubt, and hope that all of the men—in differing measures—must have felt.

"God, if You are above us, as we are told You are, look down on me and my friends and save us."[25]

Eleven

•

"Dreading to Look"

Women whose husbands or brothers failed to return this morning,
went to the morgues, dreading to look upon the dead,
in fear that loved ones might be found in the undertakers' care.
—*BUTTE DAILY POST,* JUNE 9, 1917

On the morning before the outbreak of the North Butte fire, Madge Duggan asked her husband, Manus, to stay home that day from the mines. "I asked him to stay home with me, as I felt sick." They talked about it for a while but ultimately decided that Manus could do little to help her.[1] Besides, they needed the money. There was, after all, a reason that Madge felt ill. In June 1917, Madge Duggan was nine months pregnant with their first child.

Madge went to bed that night before the outbreak of the fire. She roused briefly around 1:30 A.M. and remembered thinking that she heard Manus's footsteps, then drifted back to sleep. The next morning, Madge awoke to find relatives at her door, and it was then that she learned there was a fire at the mines.[2] Suddenly her whole world focused to a single, desperate question: *Is Manus okay?*

All over Butte, families and friends of miners were awakening to the same sickening realization as Madge Duggan. Those who lived in proximity to the mines would have heard and seen the calamity up close. Many others learned by word of mouth, as telephones and people on foot spread the news.

The first newspaper to hit the street with a partial list of the dead and missing was the *Butte Miner,* which went to press at 4:50 A.M. with a headline "Big Diaster [*sic*] at N. Butte, Many Lost."[3]

The *Anaconda Standard* was already going to press when news of the fire came through. It published a short front-page article stating that the "mine office early this morning said they did not think there is any loss of life, although it is impossible to ascertain what conditions are below the 2,400-foot level." Within a few hours the *Standard* issued its first "extra," conveying a more accurate description of the fire along with a preliminary list of the dead and missing. When the North Butte Mining Company released its first official list of casualties, the *Standard* issued a second extra, this time a single sheet listing the dead. Though 8,000 of the extras were printed, demand fell short by "90 percent," and establishments all over town posted the list on windows and doors.[4]

One seventeen-year-old wife of a miner heard the newsboys calling "Extra!" when she rose to cook breakfast at 6:00 A.M. She realized that her husband had not come home the night before but assumed he was working overtime, something he did commonly. "[I]t was not until a cousin came that the awful blow fell." Her husband was later found dead.[5]

Though the newspapers provided as much information as they could, few families in those early hours were able to learn what they cared about most—the safety of their relatives. Desperate for information, they set out to search for themselves. As the morning unfolded, a flood of relatives, friends, and onlookers swelled up the Hill, "thousands of people." At the Speculator, family members "stormed the gates in a demand for names of the dead and missing."[6]

The watchman (along with soldiers of Company F) blocked the

human tide from entering the mine yard, but timekeepers Grover Mc-Donald and Thomas O'Keefe did their best to provide as much information as they could. At the gate, the watchman collected names, passing them to McDonald and O'Keefe. The two timekeepers then ran back to the office to check on the status of each man—either "up on the surface" or "still below." In those early morning hours on Saturday, few bodies had been brought up from the depths. That would change quickly as the day progressed, and the timekeepers would add a third status to their reports: "killed."[7]

Depending on the news that family members received, reactions ranged from the "delirium of joy" to "wild ravings at the shock." Some women fainted and had to be carried to nearby houses. One woman, described by the *Anaconda Standard* as "perhaps sixty years of age," stood outside the Speculator gate and prayed aloud. "Oh God, save my husband, save my two sons." The *Standard* did not list the woman's name, so it is not clear whether her husband and sons survived. According to the best list of those killed in the North Butte disaster, there is only one instance in which *three* victims shared a last name—the common name of Murphy. There are eight instances in which *two* victims shared the same last name: Borden, Cavalla, Dougherty, Erickson, Johnson, Morris, Smith, and Thomas.[8]

Because little could be seen from the gate itself, many family members pulled back to surrounding vantage points. "Every pile of lumber, eminence of ground and top of buildings was topped with a swarm of men." From a distance, the families could discern the helmet men as they surfaced, often bearing a slumped form across their shoulders. With each emergence, "the group of women made heartbreaking suppositions as to whom it might be."[9]

Over the course of that first day, some families would return home to find their loved ones waiting inside, having staggered home after making their escape from the burning mine. Scores of families, though, would find their houses empty. "[O]n their faces was written keen anxiety. Wives and mothers of missing miners awaiting their return at home became hysterical as the day wore on and still no news of their loved ones came."[10]

For family members of Butte miners, the front-seat horror of the North Butte disaster was palpable. For all of their lives, they would remember the sights and sentiments of *being there*. Family members outside of Butte, however, faced something equally horrible— the exasperating impotence of *being away*.

News of the disaster spread quickly and widely. Telegraphs and telephones connected Butte to newspapers around the nation, including the *New York Times,* the *Wall Street Journal,* and the *Washington Post.* The *Post*'s front-page coverage ran under the headline "215 Dead in Mine Fire."[11] A woman named Margaret O'Neil wrote to North Butte officials to inquire about a cousin after reading about the disaster in the *Oakland Tribune.*[12] Relatives in small towns read the news too, as did families in the dozens of foreign countries that had sent their native sons to the copper mines of Butte.

Like the family members in Butte, these far-flung families felt the same desperate fear, the same desire to *know*, the same urge to *do something*. But what could they do, hundreds or thousands of miles away? In many cases, the only answer was to write, telephone, or telegraph Butte. By noon on the first day after the fire, "telegrams began pouring into the offices of the North Butte Mining Company, city officials, the sheriff's office," and even newspapers. In the days that would follow, thousands of inquiries would come from "almost every state in the union" and from overseas. "Telephone girls" and "messenger boys" virtually drowned in the traffic.[13]

A mother from tiny Augusta, Kansas, identifying herself as "Mrs.

B. Thompson," sent a telegraph on Saturday, June 9—the first of at least six correspondences over the next two months: "WIRE ME IF SON VERNON THOMPSON SAFE OR NOT." The next day, June 10, Mrs. Thompson sent another telegraph, this time with a short physical description of her son. On June 11 she received a reply from the North Butte. "VERNON THOMPSON STILL AMONG THE MISSING ARE MAKING UTMOST ENDEAVORS TO DISCOVER MEN."[14]

Like Mrs. Thompson, many of the out-of-town relatives sent along physical descriptions and even photographs of their loved ones. One wife devoted fifty-one words to the description of her husband's large, "dangling mole."[15] Another woman wrote in search of a miner named Lewis Cartright with a "dimple in his chin" and teeth "small and all even."[16]

In some instances, out-of-town friends and relatives traveled to Butte in order to search for their loved ones. When a list of the dead and missing was published in the small town of Roberts, Idaho, local residents were "shocked to find the name of Godfrey Galia among the missing." His fellow lodge members in the International Order of Odd Fellows, having read about the low likelihood of finding any more survivors, elected a representative to travel to Butte for the purpose of recovering Galia's body. L. V. Ledvina—who also served as the local postmaster—was the chosen member and departed promptly for Butte.

In Butte, meanwhile, family members who could not determine the status of a missing miner at the mine itself shifted their searches to the hospitals and funeral homes. Crowds outside of Butte's three hospitals were so thick that a passage had to be cleared when a new injured man arrived. The crowds ultimately grew so large that the ability of the hospitals to operate was jeopardized, and it became necessary for the police to cordon off the entryways.[17]

The hospital staff was as sensitive as possible to the ordeal of the waiting family members. As soon as the identity of a wounded miner was known, a nurse would appear at the window and paste up a slip of paper with the man's name. Members of the crowd would press forward to view it. A lucky few would discover that their loved one was inside with only slight injuries. For most, though, the result was simply a continuation of the aching uncertainty.

Still, at least the hospitals offered the *hope* of a happy result. For trips to the undertaking establishments, by contrast, the prospects ranged narrowly between *bad* and *worse*. For those who failed to find their relatives and friends, the uncertainty would continue. Worse by far was the dread—and for many the reality—that "loved ones might be found in the undertakers' care."[18]

The scene was the same in all six of Butte's funeral parlors (and later at a morgue established at the Speculator). Long lines formed outside each establishment. Inside, people shuffled silently past the bodies, hats removed in a gesture of respect. "Sighs, sobs and tears were prevalent, and occasionally a scream, as some poor mother, wife or sister or daughter identified the body of her loved one."[19]

For days after the fire, family members trudged a grim path among the undertaking establishments, scanning the long rows of unidentified dead. Salvation Army officers, making the rounds of victims' families, reported that often no one could be found at home. "This was explained by the great stream of relatives constantly flowing back and forward between the undertakers' room, the hospitals and the mines."[20]

In some families, circumstances required mere children to make the awful rounds. A Polish boy, "perhaps seven years old," showed up at one of the hospitals and asked the nurse in charge, "Is my papa here?" He had already walked to all of the other hospitals and funeral homes. "Mama is sick in bed and couldn't come, so I'm looking for him," explained the boy. "We are afraid he was killed." The hospital could say only that the boy's father had not been admitted.[21]

Many of the bodies were horribly disfigured, so that "it is with difficulty that the closest friends can tell them, so changed are they from their appearance in life."[22] By the second day after the start of the fire, the heat in the mines had begun to take a grisly toll. "Some of the bodies have become so swollen that they are twice their normal size and the skin in many instances has split under the tension of the distended forms."[23] When faces could not be distinguished, identification was sometimes based on the smallest of details. One funeral parlor notified the newspaper of an unidentified body with a "thumb nail on the right hand split way back."[24]

The deteriorating state of the bodies resulted in numerous erroneous reports about who had been killed. A Butte man named George Bigcraft was told by several people that the body of his brother had been seen at one of the local funeral homes. George rushed to the establishment and with great relief discovered that the body in question clearly lacked two of his brother's distinguishing characteristics: a "scar on the lip and a broken little finger."[25] Sadly, George Bigcraft's relief would be short-lived. A few days later, the body of his brother would be found.

Some bodies could be identified only through the contents found in pockets, such as notes, watches, and keys. Jewelry sometimes offered another clue, though in some cases the fingers had bloated to such a degree that rings could not be seen.[26]

By Sunday, June 10—the second day after the start of the fire—health concerns about the large number of dead bodies would force Butte officials to establish a public morgue at the Speculator. By necessity, the process for family members to search the dead became more controlled. Coroner Aeneas Lane instructed family members to "pass through the morgue in single file," and to "keep moving so as to expedite the work of identifying the bodies as much as possible, as decomposition has ensued to such an extent in many cases as to render hurried burial imperative."[27]

Viewing hours and time limits were established. Amid the families' anguish and uncertainty, newspaper articles informed them that "relatives who have not found their dead may look at these [new bodies] after 9 o'clock this morning and until the removal at 4 P.M." The "removal" was a euphemism for mass burial. Families that *did* find their loved ones at the Speculator morgue were told that they could "stipulate the burial plot, but the body must be taken to the cemetery at once."[28]

Scores of bodies were buried before they could be identified. Where possible, the contents of dead men's pockets were preserved in an envelope corresponding to the unmarked grave. If relatives later recognized these personal possessions, the body could be reinterred. Many of these anonymous graves were decorated with flowers sent from nearby Missoula, the "Garden City," whose citizens sent more than a ton of flowers via a caravan of automobiles.[29]

B utte responded to the emerging disaster with an outpouring of generosity and emotion that is characteristic of the town to this day. Mining was the city's lifeblood, and almost every citizen felt the impact of the disaster through some personal connection. "Those who had not lost personal friends—and they were few in number—sympathized with those who had."[30]

In the days that followed the outbreak of the fire, Butte had the shell-shocked feel of a town under siege. Flags flew at half-mast; crowds gathered outside of newspaper offices to read the latest postings; throngs of people huddled on street corners to exchange news; and everywhere people "walked as if they feared to tread too heavily."[31]

Businesses large and small offered special accommodations for the families of victims. Hennessey's Department Store, the flagship of downtown Butte, ran an advertisement in the Sunday papers. "To those who have lost either friends or relatives in the mine disaster, we realize there is nothing we can say or do to lighten the burden of your grief." Though the store normally closed on Sundays, the ad offered to "take care of emergency needs for apparel today" and listed the store's four-digit phone number.[32]

Conover & Spillum, a billiard hall, conducted a "relief fund benefit," donating to victims' families the total receipts from its pool tables as well as 10 percent of "receipts from sale of tobaccos, cigars, cigarettes, fruits, ice cream, soft drinks and all merchandise for one day." Grocery stores offered liberal credit. The Finlen Hotel announced that there would be no music in its café, an example soon followed by Butte's other restaurants. Baseball games were canceled and the amusement park closed.[33]

Not every establishment, it should be noted, was quite so sympathetic. A businessman named C. M. Covert of Whitehall, Montana, sent a letter to the North Butte Mining Company asking about a dead miner named John P. Lalanne. The death, noted Mr. Covert, left Lalanne "owing me an account for $165." Covert asked for advice on "how to go about getting this."[34]

There was a significant organized infrastructure for aiding the families of miners—at least in those early days after the fire. One prominent organization was the Salvation Army. After communicating with mine officials to collect the most current list of the dead, Salvation Army officers paid individual visits to the families. At one level, they acted as a sort of 1917 version of grief counselors, as they were "accustomed to dealing with the sorrowful side of life." The Salvation Army officers included among their ranks former miners and one adjutant who participated in the relief work after the San Francisco earthquake. When necessary, the Salvation Army could sometimes help pay for poor families' most immediate needs, including funeral expenses. Through its Lost and Missing Friends Department, the Salvation Army also helped to locate and notify the far-flung families of dead miners.[35]

The day after the fire began, the entire membership of the Montana State Industrial Accident Board arrived in Butte. In 1915, Montana's legislature had passed a statute that fixed compensation levels for victims of industrial accidents—while also limiting the liability of the mines.[36] "[I]t is fortunate that we have a compensation law that makes it clear without quibbling just what is coming to each person's dependents," said a member of the board. "We will take up the question of settlement without delay and will also see to it that there is no suffering from a financial standpoint among the relatives of the dead."[37]

The president of the North Butte Mining Company, Thomas Cole, wired a message to Butte mayor Maloney that the company would "spare no expense in bringing relief to the bereaved." Early estimates put the dollar value of compensation at somewhere between $500,000 and $750,000—a considerable sum for the day. In reality, the corporate relief provided to families of the dead miners would turn out to be quite different from the resolute promises in the immediate aftermath of the disaster.[38]

Of all the efforts to help the families of dead and missing miners, none proved more vital than the safety net of family and friends. One Butte newspaper described it as "the wonderful comradeship of the workingmen." All over town, families and neighbors did what they could, thousands of small actions that together helped to lighten the blow. They took in children, cooked meals, managed chores, or simply listened with a sympathetic ear.[39]

M adge Duggan was among those who could rely on the comfort of family. She would stay at her parents' house throughout the long days after the outbreak of the fire—awaiting the birth of her child even as she wondered at the fate of the child's father. Manus's mother would also arrive, stepping off the train from Pennsylvania to learn that her son was missing underground.

Nothing, no one, could remove the burden of Madge's anguished wait. But her family, at least, could shoulder some small portion of the load.

Twelve

•

"NOW IS THE TIME"

Now is the time. Boys, now is the time.
We can make it if you muster all the strength you have left.
—MANUS DUGGAN, JUNE 10, 1917

At some indeterminate point—probably around midnight on Sunday, June 10—Albert Cobb found himself sitting beside Manus Duggan near the crude bulkhead wall. Approximately twenty-two hours had passed since the men had sealed themselves in. Cobb, who had listened to Ernest Sullau's plea for a water bucket in the early minutes of the fire, would now sit shoulder to shoulder with Duggan in the climactic hours behind the bulkhead.

Cobb remembered seeing Duggan writing. "Are you making out your will?" he asked, intending the question as a joke.

Duggan's response was dead serious. "Yes, Tussie," he whispered, using Cobb's nickname. "I don't think we are going to get out."[1]

"Do you have a wife?" asked Cobb.

Duggan nodded. "I have a wife and mother waiting for me."

"We're in the same fix," said Cobb.

Duggan worked for a long time on the will, perhaps as long as an hour. Murty Shea, one of the other men who was near Duggan, remembered that Manus tried to "hide his writing, hoping to keep up the courage of the others."[2]

When Manus finished, he returned the notepad to his pocket. "We did not ask him what was in it and he did not tell us," remembered Cobb. Some of the other men also wrote out wills, though most could do little more than lie on the floor, fanning the air in front of them.

The men with Duggan were running out of words to describe their suffering. Wilfred La Montague, the man scheduled to be wed the following day, said simply, "Finally the air got so bad behind the bulkhead we couldn't stand it any longer."[3]

Starkly underscoring the situation, at 1:00 A.M. the sole remaining carbide lantern would no longer burn. Strangely, there are no complaints of darkness in the various men's accounts. This fact appears to confirm that at least one of the men—John Wirta—had an electric flashlight.[4]

More significant than the issue of darkness is what the burned-out carbide lantern implied about the quality of the air. The lantern failed not for lack of fuel, which the miners had in abundance, but for lack of oxygen. The men made immediate, repeated efforts to relight a lantern or even to strike a match, but no flame would take.

At the time they sealed the bulkhead, it is estimated that the men had available to them approximately 9,000 cubic feet of pure air, of which approximately 20 percent was oxygen. The burning of carbide lanterns requires air with at least *13 percent* oxygen—meaning that the oxygen supply had dipped to below that level when the lamp flickered out.[5]

There is another important marker: A minimum of approximately *10 percent* oxygen is necessary to sustain human life—especially when other impurities are present in the air. In the case of the Duggan party, there would have been high levels of carbon dioxide constantly exhaled by the twenty-nine men. It is also likely that at least some carbon *monoxide* was present from the fire. Manus's bulkhead was tight but apparently not impermeable. The men reported smelling gas beginning only an hour after the bulkhead was sealed and at various times thereafter.[6]

The miners might have already been dead were it not for the rotation system that Duggan established in the early minutes behind the wall. According to the official *Bureau of Mines Accident Report,* "The men kept moving about while behind the bulkhead, and occupied all parts of the confined area, thus mixing the pure with the impure air and diffusing both the oxygen and carbon dioxide, which accounts for the fact that some of the men were not totally overcome."[7]

Still, by 1:00 A.M. Sunday, the men with Duggan had dipped into a perilous gray zone—still above the critical threshold of 10 percent oxygen content, but below 13 percent and falling by the minute. The miners, of course, knew nothing about the complex measurements and math. What they did know was that the failure of the lantern was a terrifying sign.

The lantern incident was likely the immediate catalyst to a major revolt against Duggan, calling again for the bulkhead to be torn down. An unnamed miner "made a speech," and though his words are not recorded, there were persuasive arguments against Duggan's strategy of staying put. Many of the men believed that death was inevitable. "Let's rush the thing and go out and die an easy death in the gas and not suffer these tortures," said one miner, possibly the main speaker.[8]

It is also possible to imagine a compelling argument that breaching the bulkhead was the miners' best chance for survival. The more time that passed, the more the terror of the known came to outweigh the terror of the unknown. The miners knew that they were suffering immensely behind the wall. They knew that there had been no rescue after almost twenty-four hours. And now, most ominously, they knew that there was not enough oxygen in their cavity to sustain the flame of a carbide lantern. Wasn't it better to make a run for it while they still had strength? At least to try?

Whatever his precise words, the miner who made the speech was a savvy politician. After stating his case, he forced the issue by calling for a show of hands. Who favored tearing down the wall? It was a dramatic moment. As Duggan watched from his position in front of the bulkhead, a majority of his fellow miners voted against him—voted to breach the wall.[9]

Duggan, though, was not ready to give up. He pointed to the bulkhead. "You boys are going to leave that there as long as we have life."[10] Manus reached for the wadded shirt that plugged the bulkhead's test hole. He pulled it out and waved it in front of the men, the terrible stench of the gas still thick in the fibers of the cloth. But Duggan did more than just appeal to their fear—he appealed to their "gameness to fight it out," to the toughness and resilience he had observed so often among his fellow miners.[11]

Manus's argument carried the day. "They wanted to break the bulkhead and get out Saturday night," remembered Willy Lucas. "But Duggan talked them out of it."[12]

"Duggan persuaded us to wait several hours," said La Montague.[13]

At this point, though, the miners were divided into two camps. Gone was the solidarity of their wild dash to the blind drift and their joint enterprise in the construction of the bulkhead. Now, twenty-four hours later, Duggan found it necessary to protect those same walls from internal attack. At least two men—Ned Heston and Albert Cobb—helped to stand guard.[14]

Josiah James described the grim scene around 2:00 A.M. on Sunday morning: "[E]very man in the prison was moaning with pain, for the air was about exhausted. Men crawled about in their agony. Prayers mingled with blasphemy; laughter, from those who were delirious, with the crying of other men."[15]

The prohibition on drinking the copper water had long since fallen by the wayside. After 2:00 A.M., in fact, the men thought it tasted good. "The only problem was the shortage of it."[16] By 4:00 A.M., remembered Josiah James, the half keg of copper water was all gone. "After that the suffering increased hourly."[17] Another man described how "my lips stuck together, I was so dry."[18] La Montague stated his bottom line: "I would have given $1,000 for a drink of water."[19]

At least one account describes a last effort by some of the miners to breach the bulkhead, reporting that once again "Duggan stood in front of the bulkhead and ordered the miners back."[20] But by 8:00 on Sunday morning—now thirty hours since the bulkhead was built—the men were too weak to do anything but try to breathe. "[W]e were all quiet except for the breathing," remembered Wirta.[21] A miner who had been humming throughout the ordeal began to choke.[22]

Wirta, another of the men in close vicinity to Duggan, would tick down the final hours on his watch. "It was simply an eternity of hell," he remembered. "Even the groans of the men as they breathed seemed something unusual and disconnected. Our eyes seemed to start from our heads, and even the strongest were losing consciousness fast."[23]

On top of their deteriorating physical condition, the mental strain of their long trial pushed the miners to the edge of breakdown. Ned Heston, one of the men helping Duggan to guard the bulkhead, thought "the men were losing their minds."[24]

Duggan knew that he could not wait much longer. At 1:00 on Sunday afternoon, Duggan pulled the shirt-plug to check the outside air. *Still gas!* But at 2:00 P.M., Duggan checked the air again, and Wirta described what happened next. "[T]he nipper told me that he was going to break through, try to make the shaft and get assistance for the men."

"We are all dying, at any rate," said Manus, "so I might as well take the chance of saving the men."[25]

After thirty-six hours behind their wall, their moment of truth had arrived.

"Now is the time," Duggan told the miners. "Boys, now is the time. We can make it if you muster all the strength you have left."[26]

Duggan reached up and began to rip down the bulkhead.

Thirteen

•

"Men Alive!"

Men alive on the 2,400!

—NEWS FLASHED "BY WORD OF MOUTH" ON THE SURFACE OF THE
SPECULATOR, SUNDAY, JUNE 10, 1917[1]

I t didn't take much for Manus Duggan to batter down the bulkhead. Once the boards were peeled away, the rest was just cloth
and dust.

When John Wirta, the man with the watch and the flashlight, saw
what the nipper was doing, he rose up to help, "as willing to die outside as in the drift." Wirta worked in a "half consciousness," though
he remembered Duggan "breaking through the bulkhead, stumbling
and falling, and then getting up and staggering away in the direction
of the shaft."[2]

Having earlier battled in opposition to the destruction of the wall,
Manus was now the first to risk the outside air. Josiah James recalled
that Duggan, at this point, "appeared to be in the best physical condition of all the men entombed."[3]

As they sucked in their first breaths outside the wall, the men
realized to their great relief that they could breathe. "The air," reported Willy Lucas, "was not so bad as it was when we first built the

bulkhead."[4] Duggan's insistence on waiting, it appeared, had been justified. The available evidence—including from helmet men who had explored in the general area of the bulkhead—suggests that an earlier attempt to breach the wall would have resulted in the deaths of all the miners.[5]

Most of the men were able to follow Duggan. "Those who were too weak to get up were assisted," said Josiah James, "and it was decided that to save their strength all hands should crawl to the station."[6] They were a *quarter of a mile* away from the 2,400 Station of the Speculator shaft. As they crawled across the dirt and gravel on their hands and knees, many of the miners were naked or nearly naked, their clothing having been sacrificed in the construction of the bulkhead. Still, "[m]ost of the men felt better when they found that they were going to make a break for the shaft," recalled Steve Bodnaruk. "Even those who had lost their right mind still had sense enough to follow [Duggan's] lead."[7]

Three of the men stayed back behind the bulkhead, including Albert Cobb. Cobb described his strategy. "My idea was, the air would be better after all the men had left and that I had a better chance in there than outside in the gas." John McGarvey and Ned Heston also stayed back, though in their cases they may have been too weak to move. "They were nearly all in," said Cobb.[8]

In the men's accounts of what happened once the bulkhead was down, no one in the Duggan party complained of darkness. It is known that Duggan did *not* have a light. Given his knowledge of the underground workings, he may have been confident that he could find the station by feeling his way along the rails.[9] It is possible that some of the men were able to relight their lanterns; the clearer air would have allowed it. It is also possible that the men were able to follow the beam from Wirta's flashlight. Finally, there is the possibility that the 2,400 Station of the Speculator cast off sufficient light to guide the men. By the afternoon of Sunday, June 10, electricity had been restored to the Speculator shaft stations. Unlike in the Granite Mountain stations, fire had not destroyed the wiring in the Spec.

Josiah James remembered that the air grew better and better as they approached the station, a fact that gave additional strength to the struggling miners.[10] With the dim light in the station came a glimmer of hope that the men might survive.

H ope, for the families on the surface, was more difficult to discern. The people of Butte woke up on the morning of Sunday, June 10, to grim headlines. "Death Exacts Enormous Toll" read the *Anaconda Standard*. "Mine Fatalities Estimated at 193" led the *Butte Miner*. The newspapers published the names of the twenty-seven identified dead along with a long list of the missing. Families scanning for the names of their loved ones in the *Butte Miner* would also have read another front-page headline: "Hope Is Faint That Any of the Missing Miners Have Escaped." The *Anaconda Standard* stated bluntly that all those missing were "probably dead."

Butte's forty-four churches were full under normal circumstances, but on that Sunday the pews were bursting. Diverse clergymen struggled to answer a difficult question: How can a loving God permit horrible disasters? At the People's Church, Reverend Lawrence A. Wilson frankly acknowledged the betrayal that he knew his parishioners must feel. "Warm assurances are given each Sunday from our pulpits that God cares for each individual," he said, "that the very hairs of our heads are all numbered, and that His love is so great that not even a sparrow falls without His love and care.

"Then comes some great disaster like that of yesterday." Reverend Wilson recalled the book of Psalms, where a taunt is thrown in the face of a believer: "Where is now thy God?"

It must have been cold comfort when the reverend reminded his congregation that "we are apt to be overwhelmed by the unusual. We note the exception and not the rule." Nor could the families of dead miners find solace in Wilson's hope that the North Butte disaster would point the way toward "the means of averting like disasters in the future." For the people of Butte, the scars were too thick. The present disaster, though Butte's worst, was hardly Butte's first. What about the eight young nippers killed in a cage accident at the Black Rock Mine in 1911? Or the sixteen dead in the 1915 powder explosion at the Granite Mountain collar? Or the twenty-one dead in the 1916 fire at the Pennsylvania? And that was just the last six years. Why hadn't those past calamities prevented this present one?

No doubt sensing that his audience remained unconvinced, Wilson saved his most persuasive argument for last: "But see the higher laws at work!" he implored. "Humanity is again vindicated. Here is a man rushing back into the gas and flame to rescue a comrade. Here are the intrepid helmet men, defying death every time they enter a shaft. Here is the great-hearted city of Butte, athrob with sympathy and tenderness."

Reverend Wilson looked out at the tired eyes staring back at him, eyes desperate to understand. "Do you not see God here?"[11]

As John Wirta remembered it, he was the "first man to reach the shaft." In his half-stupor, he managed to locate the pull cord that communicated with the surface. "I reached for the bell cord signal," he remembered. "It was a supreme effort for me to count the nine pulls accurately. I am not sure that I did."

Other men began to crawl into the 2,400 Station. Wirta recalled being "possessed with an idea that somehow we were saved." But he could not yet know. Whatever the outcome would be, his last bit of energy was spent. "I lay down beside the shaft to either be saved or to die." He probably passed out, not even able to check the watch that had helped to chronicle so much of their collective suffering. "I remember nothing more."[12]

Willy Lucas, the seventeen-year-old, was another of the miners who managed to crawl the quarter mile to the 2,400 Station of the Speculator. He too pulled the bell cord. Then he waited.

A Speculator hoist operator named Dean Selfridge was sitting at the controls when the bell in the engine house suddenly began to ring. *One . . . two . . . three . . .* Whether Wirta or Lucas managed to send the accurate emergency signal of nine bells is not clear. Whatever the actual number, though, Selfridge got the message.[13]

It was a stunning turn of events. When the hoistman spread the word, others on the surface at first refused to believe.[14] "It was like a voice from the grave to hear the signal."[15] Skeptical or not, the rescuers now leapt into action. "Extraordinary activity seized the men who had been cast under the spell of the presence of death." Helmet men pulled on their equipment to make ready for a descent. Cots and tables were brought back out, and ambulances were called. The telephone company was alerted and asked to call the doctors *back* to the mine. Believing by Sunday that there was no longer a chance of finding survivors, the physicians had long since been sent home.[16]

Within minutes, hoistman Selfridge was back at the engine room controls, lowering two helmet men, Martin Howard and Ed Rucker, into the depths. On the surface, meanwhile, a crowd of rescuers and miners began to gather around the Spec headframe, hoping for a miracle.

The men at the 2,400 Station—those who were conscious—would have seen and heard the cage coming. The rattle reverberated downward as it plunged into the earth, and the moving cables were visible through the gaping opening between the station and the shaft.

Unbridled joy broke out when the cage stopped at the station and two helmet men stepped out. The trapped miners "clamored all over the two men," then jostled for position on the cage—which could only hold five or six men. It took a while before the rescuers could convey that only a few could be transported at a time, starting with the most gravely wounded. One unidentified miner—possibly Wirta—was delirious. He became hysterical at the sight of the helmet-clad men and had to be restrained to keep him from running off.[17]

Charles Negretto, Yrja Johnson, and John Wirta were the first three men on the cage to the surface, accompanied by one of the helmet men. The other rescuer stayed back to guard the shaft, assisted by Willy Lucas. "The skip [cage] looked good to me," recalled Lucas. Still, "Rucker and I stayed there to keep the men from falling into the shaft."[18]

The crowd beneath the Speculator headframe erupted in cheering when the cage surfaced and then opened to reveal three miners—the first survivors in more than twenty-four hours. Negretto and Johnson were both able to walk off the cage. Negretto "refused assistance and walked without a waiver [*sic*] to the timekeeper's office." Johnson, according to one report, "looked as though he was just coming off the regular shift"[19] (except, presumably, for the fact that he was mostly naked).

The rescuers wrapped blankets around the men, and doctors swarmed over them. Water and food appeared on outstretched arms. The rescued men laughed and cried, embracing those who had delivered them from the depths. Someone offered Johnson a drink of brandy, which he declined because it was "against the rules." A mine official informed him that the rule would be waived, and Johnson took a swig.[20]

John Wirta had to be carried off the cage and was taken to the Speculator office, where doctors applied first aid. He would later be transported to the hospital, where he recovered.

Almost immediately, the cage was sent back down the shaft, this time with two doctors aboard.

Still behind the now-tattered bulkhead, Albert Cobb was beginning to reconsider his decision to stay back. He had heard nothing of the men who left with Duggan. Had they found safe passage? Certainly they had not returned. Behind the bulkhead, meanwhile, there remained only McGarvey, Heston, and himself, and "the groans from these other two men almost set me crazy."[21]

Heston kept hollering, "Oh Tussie! Oh Tussie!" But Cobb did not have the strength to reply.

After a period of time that Cobb estimated as an hour, he made the decision to take his own chances outside the bulkhead. He started to crawl out.

"Where are you going, Tussie?" asked Heston.

"With the rest," answered Cobb.

"Will you wait for me?"

In the early minutes after the fire, Cobb had risked his life to help spread the warning to other miners. Now, though, after thirty-seven hours, he was beyond his limits. "No," Cobb told Heston. "I'm lucky if I make it. Come yourself."

Cobb, though, did not go past the bulkhead. He found better air there and stopped to breathe it in. "I had been there about a half hour when I heard footsteps and knew them to be too quick for the men who had left. I sat and listened and a rescue man whistled and I let out one yell."

Heston and McGarvey began to yell too. Then they saw lights.

"When I saw the lights coming around that drift I could not help but cry," remembered Cobb, "because the suffering had been so terrible."

McGarvey was in critical condition, the worst of the three. Doctors treated all the men while they were still underground and then helmet men spirited them to the surface. McGarvey had to be hospitalized, but all three men would live.[22]

B eneath the Speculator headframe, happy scenes continued throughout Sunday afternoon. Fresh cheers greeted the skip cage each time it delivered a new handful of men to safety on the surface. After two days of watching the helmet men remove disfigured bodies, the onlookers reveled in sharing the joy of the survivors. They watched as Spiro Beyersich, described in the newspapers as a "big Austrian," sucked in a few big breaths of surface air before complaining, "Hell of a mine—I've lost two shifts." The crowd got a good laugh at that one, and someone handed him a cigar. He walked off smoking it "like such things that he had gone through were ordinary occurrences."[23]

The rescued men huddled in blankets, gulped down water and coffee, and then wolfed down food. Mike Spihr ate four sandwiches. Newspapermen snapped photos of the miners' strained but happy faces. The papers took somewhat embarrassed note of the "Old Country" greetings between the "foreigners" and the men who awaited them on the surface—presumably referring to bear hugs and cheek kissing.[24]

Nine of the rescued men would require hospitalization. A few of these may have made themselves ill by drinking gas-contaminated drinking water from a keg that they found in the 2,400 Station.[25]

In a time before radio broadcasts, no instant bulletin would spread the joyful Sunday afternoon news of the rescued miners. Officials did place telephone calls to notify any families who could be reached by that means—a status that included some but by no means all. Telegraphs were available to dispatch news out of town. Until the Monday morning newspapers, though, the rest relied on word of mouth—"Men alive on the 2,400!"—which in Butte that day seemed to spread as fast as any electric current.[26]

News traveled particularly fast to the neighborhoods surrounding the mines, where many of the miners lived. It is not known whether Madge Duggan was one of the relatives who rushed to the gates outside the mines, but certainly many did. "Women began to besiege the mine office," wrote the *Anaconda Standard*. "A city which had settled down to mourn its dead was once again revived by hope."[27] For a lucky few, the fresh hope would find consummation. For most, of course, it was just another cruel twist.

The family of Josiah James was among the lucky. The mine arranged transportation in automobiles for the miners who did not require hospital care, but many of the men, including James, insisted on walking home under their own power. He entered a house transformed in an instant "from one of mourning to one of rejoicing."[28]

The shock of seeing Josiah alive was particularly sharp for the James family, since a cousin had already identified his body at a local mortuary. There had, though, been one small shred of doubt. One of Josiah's roommates asked to look at the corpse. The roommate told the cousin that it could not be Josiah James because Josiah "had hammered toes and that man's toes were straight."[29]

In 1980, at the age of eighty-nine, Josiah James gave an interview to the *Montana Standard* (formerly the *Anaconda Standard*). Duggan, of course, had not been the only man to scratch out a will as he sat trapped behind the bulkhead. Sixty-three years after the disaster, James showed his interviewer the scrap of yellowed paper that he still carried in his billfold: "Notify Mrs. A. Laity, Port Hammond, B.C., Canada, before burial," he had written. "Have month's paycheck in pocket—$140. All of this for dear mother and what I have left. Love to all. Goodbye. Josiah, June 8, 1917."[30]

Like the family of Josiah James, other relatives and friends of the men in the Duggan party would be shocked by the resurrection of their loved ones. Postmaster L. V. Ledvina of Roberts, Idaho, was the faithful Odd Fellow whose fellow lodge members had elected him to travel to Butte and recover the body of Godfrey Galia. Upon his arrival, Ledvina went to Galia's boardinghouse, expecting to collect a dead man's possessions. Instead, Galia himself answered the door.

"We thought you were dead," said the stunned Ledvina.

"I thought so myself for a time," said Galia. "But thanks to Manus Duggan we got out alive."[31] Though alive, Galia was not unscathed. Two days after escaping the mine, his lungs and stomach still ached, and he worried that the damage might be permanent. "I won't go back to the mines anymore," said Galia. After helping Ledvina tend to the body of another Austrian friend, Galia decided to return to farming in Idaho, "where you can't get killed in such a terrible way as lots of my friends were killed."[32]

Galia's decision, though rational, was not typical of the miners who survived the tragedy. "Sure, I'm going back to work the mines, and the Speculator if I get the chance," said Steve Bodnaruk. "A man has only one time to die and that time may as well come underground as on top. It's all in the game."[33]

All afternoon, the Speculator's chippy cage shuttled its precious treasures from the 2,400 Station to the surface. The cheers continued, though as the day wore on, an uneasiness crept into the celebration. As the miners and rescuers began to sort out the confusion above and below ground, it eventually became clear that men were missing. By late afternoon, an accounting could be made for only twenty-five of the twenty-nine men who took refuge behind the bulkhead.

One of those missing was a young nipper that the survivors all praised as the man who had saved their lives. "There would have none of us escaped alive if it hadn't been for Duggan," said Murty Shea. "I am hoping and praying he is alive."[34]

Duggan's fellow miners urged a quick effort to search for him and the three other missing men, telling as much as they could remember about Manus's movements. Extra crews of helmet men were quickly dispatched, guided by whatever kernels could be gleaned from the hazy and confusing reports of the rescued men.

The official *Bureau of Mines Accident Report* stated—mistakenly— that Duggan "took the wrong direction after breaking through" the bulkhead.[35] It is clear from multiple witnesses that Duggan went *first* to the 2,400 Station—leading his fellow miners along the way. John Wirta, who stood beside Manus as they tore down the bulkhead, was one of the men who remembered following the nipper in the direction of the station.

Steve Bodnaruk was another. "I'm sure the nipper got to the shaft," he said. "I remember seeing him there." He specifically recalled Manus "signaling on the pipes" after the rope cord had been pulled.[36] One newspaper account also reports a witness (unnamed) who saw Duggan at the 2,400 Station, "talking incoherently about getting water for 'his men.' "[37]

Whatever Duggan's reasoning—concern about the men who stayed back, mistaken judgment, or simply incoherence—after reaching the 2,400 Station, he turned around and *walked away.*

By sunset on Sunday, June 10, helmet men had scoured the tunnels in the vicinity of both the 2,400 Station and the Duggan bulkhead. They found the body of Joseph McAdams—the twenty-sixth member of the Duggan party to be accounted for—but they did not find Manus Duggan or the two other men. On the surface, meanwhile, "great clouds of smoke and gas" continued to pour from the Granite Mountain shaft.[38]

Duggan's comrades began to resign themselves to the worst. Albert Cobb was one of the men who would give a lengthy interview to the Butte newspapers about his experiences underground. By then, the Duggan story had captivated all of Butte. Cobb thought back to the will he had watched Duggan write, then stuff into his pocket. "His body will be identified by that."[39]

At the end of his interview, Cobb talked about his wife's joyful reaction to his homecoming. "I knew it would end like this," she had told him. "I never quit."

Madge Duggan didn't quit either, but for her, the horrible wait would continue.

Fourteen

·

"BAMBOOZELING OR ABUSE"

Stand up in full dignity of real manhood, and do not,
under any circumstances, tolerate in the future as in the past,
from any boss any bulldozing, brow-beating,
bamboozeling or abuse of any kind.

—LEAFLETS POSTED BY STRIKERS AT THE MINES, JUNE 1914[1]

B.K. Wheeler did not aspire to a career in politics, or even show much interest at first. In Butte, though, politics seemed eventually to envelop everyone. In a grand irony, it was Anaconda that gave Wheeler his first elected office. In 1909, the Company put the name "Burton K. Wheeler" on its slate of twelve candidates for the state legislature because of concerns that its existing slate was too Irish and too Catholic. "By God," complained one party man, "you've got to have some Protestants on here."[2]

"Being neither Irish nor Catholic," said Wheeler, "I had two points in my favor."[3]

Unwittingly, Anaconda had helped to create the man who would rise as its greatest nemesis. At his first session of the state legislature, Wheeler ignored a Company line to which he was expected to adhere, then spurned efforts to buy his vote (bribery was still alive and well in Helena). Wheeler's first bill was a measure to level the playing field for workers bringing suits against their employers. The legislation attempted to abrogate the archaic "fellow servant" doctrine, which stated that an employer could not be held liable for losses to one employee that were caused by the negligence of *another* employee. Many mining accidents involved another worker's negligence, leaving miners without compensation for this common category of injury.

On a Friday afternoon, it appeared Wheeler's bill would pass, but the Company quickly mobilized its Helena lobbyists. The *Helena Independent,* an Anaconda-controlled newspaper, editorialized that Wheeler's bill was "altogether too radical and sweeping." On Monday, the bill was referred back to committee, from whence it never again emerged.[4]

Wheeler's opposition to the Company line may not have helped him enact legislation, but it won him a lot of clients. Back in Butte, aggrieved workmen streamed into his law office, particularly those who had been injured in the mines. The young lawyer's practice began to thrive.[5]

Wheeler rose to national prominence after the presidential election of 1912, when Woodrow Wilson's Democrats defeated a Republican Party splintered by Teddy Roosevelt and his renegade Bull Moose Party. Wheeler had campaigned for one of Montana's U.S. senators, Thomas Walsh. When Walsh won, he repaid the favor by urging President Wilson to appoint Wheeler as Montana's federal district attorney. At thirty-one, Burton K. Wheeler became the youngest federal DA in the country.

B.K. Wheeler's politics were shaped by the predominant current of American politics in the early twentieth century—"progressivism." Progressivism had influenced both major political parties and manifested itself through both a Republican president, Teddy Roosevelt, and a Democratic president, Woodrow Wilson. Progressives were generally moderates who sought to reform government in order to address a range of social ills, including corporate abuse, exploitation of women and children, and political corruption. Progressive achievements included child labor laws, direct election of U.S. senators, women's suffrage, and landmark environmental protections.[6]

In Montana, fresh on the heels of two decades of naked political corruption and still dominated by Anaconda, Wheeler saw plenty of places to apply a progressive political philosophy. Wheeler won significant suits against big corporations, including the Great Northern Railway. But during his five-year tenure as district attorney, he also showcased a trademark independence. He brought suit, for example, against a company organized by a man named William Rae, the treasurer of the state Democratic Party and one of Wheeler's personal friends. Several other prominent Democrats were also involved with the company.[7]

Wheeler's independence would serve him well throughout his career, but no time more so than his years as district attorney. It was a period defined by remarkable turbulence, one of those eras—cyclical in American politics—in which the center was shrinking, and the agenda was set increasingly by the extremes from both left and right.

As the Granite Mountain fire burned through its third full day, on Monday, June 11, 1917, concerns arose briefly that the flames, by then primarily contained in the shaft, might break through into the inner workings of the mine.[8] In reality, though, the gravest risk now lay not in the actual fire below ground, but in the metaphorical combustibility on the surface.

Butte, Montana, was about to blow. In the eleven years since Standard Oil completed its takeover of Butte's copper resources, multiple conflicts were unfolding both in the town and in the world around it. These conflicts—all unresolved—had piled up like kindling in a bone-dry forest.

No issue, of course, loomed larger than the war. The Butte draft riot on June 5, 1917, three days before the outbreak of the fire, had demonstrated the volatility surrounding America's recent entry into the war. But war was not the only source of division in Butte. Since the 1906 consolidation of the city's copper holdings by Standard Oil, tension between the Company and organized labor had been building toward a climactic confrontation. To understand how bad things had become by June 1917, it is first necessary to understand how good things had been, relatively speaking, before the takeover by East Coast management.

Before Standard Oil, Butte was known as "the Gibraltar of unionism"—organized labor's strongest outpost in the West. Miners formed their first union in 1878, as silver brought the young town its first taste of industrialization, and by 1885 the powerful Butte Miners' Union won a coveted "closed shop" deal with the mining companies. This meant that the companies would hire only union workers, and they formally recognized the union as their counterpart in the negotiation of wages and working conditions. The effect of a closed shop was to prevent cheap nonunion labor from undermining a union's collective bargaining strength. In 1893, the Butte Miners' Union was instrumental in forming an even more powerful, national counterpart—the Western Federation of Miners.[9]

It was not just the miners' union that flourished. Butte, it seemed, had a union for everything. Almost every trade group was unionized, including construction workers, brewers, beer wagon drivers, blacksmiths, jewelers, smeltermen, engineers, horseshoers, hackmen, and teamsters, not to mention separate unions for musicians, theatrical stage employers, and theatrical ushers. In Butte the chimney sweeps—both of them—had their own union. Even the *unions* had a union—an umbrella group known as the Silver Bow Trades and Labor Assembly.[10]

For a variety of reasons, unionism and the Butte Miners' Union flourished under the rule of all three Butte Copper Kings. Marcus Daly viewed himself as a sort of benevolent dictator, and both his hardscrabble background and his Irish lineage gave him a heartfelt connection to his miners. For William A. Clark, the backing of his workers was important to consummate his political ambitions. So too for the charismatic F. Augustus Heinze, who needed his men's unflinching support for his battles—sometimes literal—with Standard Oil.

The result, in the two decades preceding the turn of the twentieth century, was a sort of golden age of organized labor in Butte. While unions in other parts of the country struggled against management—sometimes violently—conditions in Butte were relatively favorable. Mining, of course, had always been a dangerous way to make a living, but Butte wages were generally high, and a closed shop gave miners true power. In the context of Heinze's and Clark's 1900 fight with Standard Oil, miners even won the right to an eight-hour workday. From the standpoint of Butte miners, the Copper Kings may have been robber barons—but at least they were *their* robber barons.[11]

When Fritz Heinze became the last Copper King to surrender to Standard Oil in 1906, it was the beginning of a new, darker era in Butte history. While Clark and Heinze lived in New York City luxury (by this time, Daly was dead), the miners back in Montana were left holding the bag. And the new boss, it turned out, was not at all the same as the old bosses.

Standard Oil, its eyes ever fixed on the stock market from which it made its true money, instilled a ruthless drive for efficiency. In fairness, Butte in the early twentieth century was beginning to face new challenges to its position as the world's dominant copper producer, including tough competitors in Arizona, Utah, and Chile—all of whom could apply open pit techniques. With these significant sources of competition, Standard Oil would never achieve the monopoly in copper that it had achieved in oil.[12]

There was one area, however, in which the Company was unquestionably dominant. As a *supplier of jobs* in Butte and, indeed, the whole state of Montana, Anaconda had near invincible power. This power of the Company was on brazen display as early as 1903 in its climactic battle with Fritz Heinze, when Standard Oil shut down the state and forced a special session of the legislature to act at its bidding.

Anaconda used its power as a near-monopoly employer to set stark new terms for the Company's relationship with organized labor. The relentless push for efficiency led to a sharp decline in mine safety. Yet even as the hazards to miners increased, the Company fought bitterly against efforts to pass a workmen's compensation law.[13]

The Company used mass political firings to send a clear message of intolerance for radical politics. In 1903, it fired several hundred socialists and suspected socialists at the Anaconda smelter. An even larger firing took place after Butte actually elected a self-avowed socialist mayor in 1911. On the eve of the 1912 election, the Company fired 500 miners, mostly Finnish, because of suspected socialist leanings.[14] A steady stream of new immigrant labor from southern and eastern Europe meant there was no risk that the Company would be left shorthanded.

More flagrant action was still to come. In December 1912, the Company established a new system for hiring miners. Anyone applying to work at any mine (even the few mines—like the North Butte properties—not owned by the Company) was required to apply for something called a "rustling card." Only with the card could a miner "rustle" around in search of a job. Before a card was issued, the Company completed an investigation of the prospective employee, and it could deny a rustling card for any reason at its sole discretion. It was, in essence, a blacklist. Rustling cards could be denied, for example, to men who complained about safety conditions in the mines. The Company had already demonstrated its attitude concerning miners with leftist-leaning political sympathies.[15]

The creation of the rustling card system marked the dramatic end, after almost three decades, of the closed shop in Butte. No longer would membership in the Butte Miners' Union be the criterion for job selection. Now, union members could be denied a rustling card. Even more significant, cards could be given to miners who *did not belong* to the union.[16]

According to one account, the imposition of the new system forced "hundreds of men" to leave town.[17] Even for those men granted cards, the rustling card system had a tangible chilling effect. "The actual discrimination is perhaps less important than the fear and apprehension which its critics believe the rustling card instills into the minds of the miners. They feel that it is a club held over them and they fear the consequences if they speak out about unsatisfactory conditions or about the card itself."[18]

Where was the union during all of these flagrant challenges to its power and authority? At the time when its members needed it most, the once-potent Butte Miners' Union ceded the field. In the words of one contemporaneous analysis, "While the employers have been organizing, the employees have been disorganizing."[19] The radical leftists, growing in influence with every new unanswered Company provocation, were even more blunt: "Late developments in Butte, Montana, have given plausibility to the general belief that many officials of the Western Federation of Miners are simply the tools of the Copper Trust."[20]

Though most Butte miners did not go as far as the radicals on the left, by 1914 frustrated rank-and-file members could point to a long string of union failures. In both 1907 and 1912, union leaders endorsed contracts viewed as highly unfavorable to the miners.[21] Most significantly, the contracts contained "sliding scale" provisions that pegged wages to the price of copper. In 1913, miners were earning the same wage—$3.50 a day—that their grandfathers had earned in

1878. Inflation, meanwhile, had caused the cost of living to spike dramatically. The new contracts were also silent on health and safety issues, a major concern of most miners.

While a majority of union members continued to vote in support of the more conservative union leadership, a significant faction was beginning to emerge on the left.[22] Widening the division still further was the weak response of the Butte Miners' Union to the Company's string of provocations, including the mass firings of union members. The 1912 preelection firing also split the union on ethnic lines, because most of those sacked were Finns. When the Butte Miners' Union voted down a retaliatory strike, Finns felt that the Irish and Cornish members had betrayed them. For the rest of the decade, Finns would be at the core of the most radical elements in Butte.[23]

For many rank-and-file members, the last straw would be the union's failure to endorse a strike in response to the Company's establishment of the rustling card system. In the pivotal year of 1914, accumulated labor tensions would finally explode into violence.

The official report of the Montana Department of Labor and Industry would claim that the violence of 1914 "did not originate from any direct grievance against the mining companies of that district, but was the result of factional differences in the Miners' Union alone."[24] This is not an accurate description, since the factionalism came about as a result of miners' dissatisfaction with the union's failure to respond *to provocations by the Company*.

Ironically, the immediate spark to the 1914 violence would originate at the same mine—the Speculator—where so much of the death would take place in the fire of 1917. On June 11, 1914, a committee of the Butte Miners' Union showed up at the Speculator to inspect union membership cards. In a protest against the union, a miner named Michael "Muckie" McDonald refused to show his card. The next day an entire shift—400 men—refused to show their cards, and that night 2,000 miners attended a meeting to rally against the Butte Miners' Union. The insurgent miners decided to form a new union and to boycott the "Miners' Union Day" parade—scheduled for the following day and normally a major Butte event.[25]

The next day, only a few hundred miners bothered to march in the parade, and its leaders were attacked by an angry mob. The mob proceeded to march to the Miners' Union Hall and then sacked it—breaking windows, destroying furniture, and dynamiting the safe. Nine days later, more than 4,000 miners joined a new union, the "Butte Mine Workers' Union," headed by Muckie McDonald.[26]

Eleven days later, Charles Moyer, president of the old union's national counterpart (the Western Federation of Miners), rushed to Butte in a belated effort to turn the situation around. Moyer attempted to hold a meeting at the Miners' Union Hall, which was quickly surrounded by "several hundred members of the new union." Rocks and eggs were thrown at the hall, and at some point, a shot from inside the building killed a man who was walking up the steps (ironically, a member of the old union who was attempting to attend the meeting).[27]

Now complete chaos ensued. More shots were fired from both sides, wounding several and killing another man, this time an innocent bystander.[28] As President Moyer and the besieged members of the old Butte Miners' Union made a hasty escape out the back exit, a group of rioters marched up to the Stewart, a nearby mine. There, they forced the hoist operator—at gunpoint—to lower them into the mine, emerging a few minutes later with a large quantity of dynamite. Then it was back to the Union Hall.

Over the next several hours, with no opposition by local police, the renegade miners proceeded to set off twenty-six charges of dynamite. When the smoke cleared, the Union Hall had been reduced to a pile of rubble. Also in shambles, though not yet so obviously, was unionism in Butte. Divided against themselves, the miners would be no match for the concerted power of the Company.

The drama stretched throughout the summer of 1914. Only five days after the destruction of the Union Hall came the assassination of the Austro-Hungarian archduke; by the end of July, most of the world was at war. The initial effect of the war was to depress world demand for copper (only later would the war create a boom). By mid-August, half the mines in Butte were closed, putting thousands of miners on the street. Efforts to enlist members for the new union intensified, and five thousand men signed up. In some instances, miners who refused to join the union faced threats of violence or forced deportation from the town. Frequent "street demonstrations" broke out.[29]

Leaders of the new union, meanwhile, sought recognition from the Company. They also presented a list of demands, including "abolishing the system of blasting at dinner hour, and dampening and laying of the dust." Miners were told not to tolerate "bulldozing, brow-beating, bamboozling or abuse of any kind," and were encouraged to bring any grievances to the union.[30]

On the night of August 30, 1914, a rustling card house (where the hated cards were issued) was dynamited. Anaconda offered a $10,000 reward for information on the perpetrators. The new Butte Mine Workers' Union, meanwhile, maintained that the Company itself had blown up the building in order to justify occupation of the town by the National Guard.[31]

Who initiated the violence will probably never be known, but the result was clear. Two days later, the governor of Montana, "believing that a state of anarchy existed," declared martial law in Butte and sent in the National Guard. Five hundred troops armed with gatling and machine guns would occupy the town, quickly assuming control. Military tribunals were established to try civilians, including several labor leaders. Muckie McDonald, president of the new union, was convicted of kidnapping (based on the incidents of forced deportations) and sent to prison. Butte's socialist mayor was impeached for his failure to maintain order. Press censorship was imposed and public assembly forbidden.[32]

Under the cover of military occupation, the Company too took action, formally derecognizing both the new union and the old one.[33]

Long gone were the days of Butte as the Gibraltar of unionism. After 1914, the miners stood alone against the awesome power of the Company. For three years unionism would lie dormant, until the North Butte disaster lit a fire that Anaconda could not easily squelch.

•

Fifteen

"Too Good Miners"

We built it too well. We were too good miners.
—CLARENCE MARTHEY, JUNE 11, 1917[1]

On the afternoon of Sunday, June 10—at the very moment that the twenty-five survivors of the Duggan party were being lifted to safety—a plain-faced helmet man named Tom La Martine was searching a drift at the 2,200 level of the Speculator. He was looking for bodies, not men, having no knowledge yet of the Duggan bulkhead, some twenty stories below.[2]

As La Martine peered through the steamy glass of his Fleuss mask, he saw a bulkhead at the entrance to a drift. The quality of the bulkhead's construction was impressive—sawed planks nailed to crossbeam timbers and then covered with canvas. It didn't seem like a structure that had been thrown up in a panic. He rapped on one of the timbers.

No reply.

The discovery made no particular impression on La Martine. At the time he descended the mine, it had been thirty-six hours since the start of the fire. No one believed that men could still be alive, and the thought of survivors was far from his mind. He moved on. By the time La Martine returned to the surface, the thrilling pandemonium of "men alive on the 2,400" was in full swing. La Martine made no mention of the bulkhead he had seen at the 2,200.

There is no clear account for why La Martine failed to report the bulkhead on Sunday. The most plausible explanation is that after returning to the surface, he immediately became swept up in the ongoing rescue of survivors from the Duggan bulkhead. The *Anaconda Standard* reported that "[i]n the excitement attendant upon the rescue of the 25 men and upon fatigue resultant upon *his assistance in bringing the men to the surface,* La Martine did not have a chance to tell of his discovery until [Monday] morning."[3]

Whatever the explanation, it was not until the next day, Monday, June 11, that La Martine reported what he had seen. At first, his fellow rescuers were dismissive, but as La Martine thought about it, he remembered an additional, critical detail: "The bulkhead was thrown up from the *blind side.*" The other men took notice now, reporting the bulkhead to General Manager Norman Braley in the mine offices. Braley quickly produced a map showing the area that La Martine had explored. La Martine pointed to the precise location where he had seen the bulkhead. The map showed nothing.

"If there was a bulkhead there," said General Manager Braley, "it [was] built since the fire."[4]

Suddenly and fully aware of what La Martine had seen the day before, the rescuers now leapt into action. La Martine and three other helmet men threw on their equipment and were lowered down the Speculator shaft. La Martine led the others to the bulkhead, banging on the wall. "Hello in there!"

This time there came a weak reply: "There are ten men in here."[5]

Courtesy of the World Museum of Mining

"THE RICHEST HILL ON EARTH."
This partial panorama of the Butte hill shows
the dense clustering of mines.

Courtesy of the World Museum of Mining

GRANITE MOUNTAIN AND SPECULATOR SURFACE PLANTS OF NORTH BUTTE MINING COMPANY

THE NORTH BUTTE PROPERTIES.
Panoramas of the Granite Mountain and Speculator mines.
Approximately 800 feet separates the two gallows frames.

WILLIAM A. CLARK

"DEFINING DEVIANCY DOWN."
Copper kings William A. Clark,
Marcus Daly, and F. Augustus
Heinze. The battles of these
three men—and Standard Oil—
set the political tone for
the decades to follow.

MARCUS DALY

F. AUGUSTUS HEINZE

Clark and Daly photos courtesy of the Montana Historical Society; Heinze photo courtesy of the World Museum of Mining

Little photo courtesy of the Butte-Silver Bow Public Archives; Wheeler and Campbell photos courtesy of the Montana Historical Society

"GASOLINE ON FIRE."
With provocations from left
and right, Wheeler struggled
to help Montana keep its
keel in the aftermath of
the North Butte disaster.

FRANK LITTLE,
executive chairman of
the Industrial Workers
of the World

BURTON K. WHEELER,
federal district attorney
for Montana

WILL CAMPBELL,
firebrand editor of the
Helena Independent.

Courtesy of the World Museum of Mining

WHICH WAY TO SAFETY?
Miners seeking to escape the Granite Mountain and
Speculator mines faced a bewildering underground maze.
Note the water keg, the compressed air hoses,
and the absence of signage.

Courtesy of the World Museum of Mining

Skinner at the Speculator Mine

CRAMPED QUARTERS.
A skinner and his horse in a crosscut of the Speculator.

Courtesy of the Butte-Silver Bow Public Archives

MANUS DUGGAN.
Madge Duggan gave this, her only photo of her husband, to a local newspaper. The original showed a twelve-year-old Duggan in a bathing suit, but was cropped and doctored to show him in a suit and tie.

Courtesy of the United States Department of the Interior

"TOO GOOD MINERS."
A man stands behind the remnants of the J. D. Moore bulkhead, behind which ten men sought to save their lives.

Courtesy of the Montana Historical Society

WESTERN UNION TELEGRAM

Form 1201

CLASS OF SERVICE	SYMBOL
Day Message	
Day Letter	Blue
Night Message	Nite
Night Letter	N L

If none of these three symbols appears after the check (number of words) this is a day message. Otherwise its character is indicated by the symbol appearing after the check.

NEWCOMB CARLTON, PRESIDENT   GEORGE W. E. ATKINS, FIRST VICE-PRESIDENT

CLASS OF SERVICE	SYMBOL
Day Message	
Day Letter	Blue
Night Message	Nite
Night Letter	N L

If none of these three symbols appears after the check (number of words) this is a day message. Otherwise its character is indicated by the symbol appearing after the check.

RECEIVED AT 16 EAST BROADWAY, BUTTE, MONT.

117D DA 10

SANJOSE CALIF 634P JUNE 9 1917

SUPT SPECULATOR MINE

BUTTE MONT

IS JAMES D MOORE ALL RIGHT ANSWER IMMEDIATELY MY EXPENSE

MRS M D MURPHY

8PM

TELEPHONED TO
TIME
BY
HOW DELIVERED

WESTERN UNION TELEGRAM

Form 1201

CLASS OF SERVICE	SYMBOL
Day Message	
Day Letter	Blue
Night Message	Nite
Night Letter	N L

If none of these three symbols appears after the check (number of words) this is a day message. Otherwise its character is indicated by the symbol appearing after the check.

NEWCOMB CARLTON, PRESIDENT   GEORGE W. E. ATKINS, FIRST VICE-PRESIDENT

CLASS OF SERVICE	SYMBOL
Day Message	
Day Letter	Blue
Night Message	Nite
Night Letter	N L

If none of these three symbols appears after the check (number of words) this is a day message. Otherwise its character is indicated by the symbol appearing after the check.

RECEIVED AT 16 EAST BROADWAY, BUTTE, MONT.

62SX HS 50 BLUE 6 EXTRA

SANJOSE CALIF 215P JUNE 14 1917

SUPT SPECULATOR MINE

BUTTE MONT

IF MY SON JAS D MOORE LEFT LETTERS FOR MOTHER SISTER LEE OR BEN KINDLY

SEND ME COPIES OF SAME BY MAIL LET ME KNOW COST OF SAME AND WILL

SEND YOU MONEY IF HE LEFT THE ABOVE LETTERS WIRE YES AT MY EXPENSE

MRS M D MURPHY 895 SOUTH SECOND STREET

405PM

"TO SALVE THE WOUND."
Two telegrams to the "Supt. Speculator Mine"
from Mrs. M. D. Murphy, mother of J. D. Moore.

END OF THE LINE.
The note pinned
on Frank Little
by vigilantes.

BUTTE TODAY.
Note the new
water treatment
facility, lower left.
The Speculator and
Granite Mountain
mines are on the
far right of the pit.

Courtesy of the Butte–Silver Bow Public Archives

Courtesy of Sophie Silk Punke

Two and a half days earlier, just after midnight on Friday, June 9, a young miner named Clarence Marthey was working with a crew between the 2,000 and 2,200 levels. "The first thing I knew of the fire was when James Moore yelled to us that gas was breaking on the 2,000 level." Moore told Marthey and his crewmates that they "could not make the Granite Mountain shaft" and then ordered them to go to the 2,200 level.[6]

James "J.D." Moore was a shift boss, the type of man who expected a quick response to his commands. Prior to working in Butte, Moore had been a mine foreman in Virginia City, Nevada. "They called him 'Tincan Jimmy,' " recalled a miner who worked for him in Nevada, "because he used to 'can' so many men . . . he was very quick tempered." Still, Moore was respected and popular—perhaps because his frequent firings never stuck. "He never remained angry for any length of time."[7]

As Clarence Marthey remembered it, J.D. Moore's temper was on full display in the moments after he gave the order to run for the drift. Several of the men with Marthey—in a full state of panic—argued for running toward the Granite Mountain shaft. It was there, after all, that they had been lowered into the mine; it seemed like the most logical way out. By Marthey's account, "When some wanted to go to the station Moore said sharply there was no chance and that his orders were not to go to the Granite Mountain but to the 'deadend' and bulkhead."[8] A few of the men still resisted, leading J.D. Moore to settle the issue by grabbing them and pushing them down the manway.[9]

Moore gave Marthey additional orders. "Take what men you can find to the 2,200 level," he said, describing a particular blind drift. "Get all the water you can find on the way and drag what timber you see." Clarence Marthey was a young husband with two small children. As an inexperienced miner in the midst of sudden crisis, he was happy to follow Moore's confident lead. "Moore disappeared, searching for more men, but came back to supervise the building of the bulk-head."[10]

J.D. Moore's confidence rested on a solid foundation of experi-ence, including experience under fire—literally. The man who had worked with Moore in Virginia City remembered that "a fire broke out in the Ward shaft, where [Moore] was foreman. His conduct was just as brave in that instance as it was in this . . . he savied [sic] fires underground."[11]

The location that Moore chose for his bulkhead was approxi-mately 3,000 feet from the Speculator shaft. As Moore's men contem-plated the full implications of entombing themselves behind a bulkhead, some of the men again balked. And once again, Moore pre-vailed through a combination of confidence and bullying. "Some of the men rebelled and didn't want to wall themselves in," remembered Martin Garrity, a popular, chatty miner. "But Moore made them."[12]

The thick stench of poisonous fumes helped to underscore Moore's case. Compared to Duggan's group, the men with J.D. Moore fought an earlier, more intimate battle against the gas. When the men with Duggan reached his blind drift, the air initially was clear, and it was not until the bulkhead was nearly complete that deadly fumes caught up with them. With Moore and his men, by contrast, the gas was in near pursuit.

In this close-quarters combat, the men stumbled across a weapon that probably saved their lives. Near the face of the blind drift, they found an abandoned Leyner drill—including the compressed air hose that powered it. Miners with Moore reported that "smoke and gas fol-lowed them into the crosscut to such an extent that they were able to complete the bulkhead only by stationing one man blowing air from a hose to keep back the gas."[13]

The importance of the air hose was underscored when the men, nearing completion of the bulkhead, needed more material to plug holes. Someone remembered seeing a piece of canvas only seventy-five feet away and made an attempt to retrieve it. Outside of the immediate range of the air hose, though, the gas was overpowering, forcing a quick retreat.

Despite the close presence of the gas, Moore was deliberate in the construction of his bulkhead. Whereas Duggan's wall went up in less than an hour, Moore and his men took "several hours." In the Moore party's dash through the mine, they had better luck than the Duggan party in collecting tools, including "picks, saws, hammers, axes, boxes of wedges . . . also some boards from the trolley guard and nails." Like Duggan, Moore also tore down a substantial quantity of canvas pipe.[14]

Moore had one building material available to him that Duggan did not—a material that turned out to be both a blessing and a curse. It was mud. Moore's basic concept for the bulkhead wall was similar to Duggan's. He directed his men to build a two-sided wooden brattice. The ready supply of mud provided the perfect material for filling the three-inch space in between, though members of the Moore party would still end up contributing a good deal of their clothing.[15] (Stanley Lazykeics, another miner who came to Butte via Eastern Europe and then Idaho, would be naked before the wall was completed.[16])

While the mud was valuable as an airtight insulator, it also pointed to one of the important problems with the site of Moore's bulkhead. Compared to Duggan's relatively dry location, Moore's drift was damp and humid—a malarial 98 percent on top of a base temperature of eighty-five degrees.[17]

J.D. Moore gave one last order before the bulkhead was sealed, "to get the air line inside."[18] The hose, which pumped fresh oxygen from the surface, was an invaluable final addition to their last line of defense.

There were ten of them—Moore plus nine others.[19] The time was approximately 4:00 A.M. on Saturday, June 9. The Granite Mountain fire had been burning for four hours.

One mistake that J.D. Moore did not make was the promiscuous use of lights. Some of the men had carbide lanterns, though they apparently were not used at first. Shortly after closing themselves in, Moore directed an inventory of their collective supplies. A man named Bob Truax offered up a couple of candles. "It was against the rules to use candles," remembered Martin Garrity, "and Moore jumped on him at first for having them." Moore's irritation, though, passed quickly. "Afterward he praised him," said Garrity, "for they were the only light we had and we used them to test and see if any gas were coming anywhere."[20]

The candlelight inspection revealed an open pipe that seemed to be pumping gas into the chamber. It was quickly plugged.

The water supply, at least, was not among their problems. Moore delegated the responsibility for water to Clarence Marthey with the directive to "see each got his share and make it last."[21] They had three kegs, an ample amount for ten men. "When we were running we grabbed every keg of water in sight and took them with us," remembered Garrity. "We had a good supply of water and that helped out."[22]

The air, at first, also seemed more than sufficient. The size of the chamber behind Moore's bulkhead was smaller than Duggan's, but Duggan had almost three times as many men. Under normal circumstances, the amount of oxygen inside Moore's chamber should have been more than adequate for the fifty-plus hours during which the men were there.[23] In addition to the oxygen behind the bulkhead, the Moore party also had the advantage of the compressed air hose, pumping fresh air from the surface. This advantage, though, was about to evaporate.

Moore and his men had barely sealed the bulkhead when a massive, terrifying rumble shook the mine. A large portion of the Granite

Mountain shaft—from the 1,000 level down—was collapsing. The shaft contained the main air pipe—through which Moore's air hose connected to the surface. With its air supply squeezed, the hose partially failed.[24] It was a devastating blow. As a practical matter, of course, the air hose would have supplied fresh air. But the hose also represented a symbolic tie to the surface, an umbilical cord connecting the trapped men to safety above ground. As Clarence Marthey put it, "When that was cut off our spirits dropped."[25]

Perhaps reflecting the gut-punch trauma of the diminished air supply, J.D. Moore turned pessimistic about their chances for survival. Over a period of five hours—beginning shortly after the completion of the bulkhead around 4:00 A.M.—Moore wrote a series of four bleak notes to his wife, later published in the *Butte Miner*.[26]

The first note, dated "6-8-17" and probably written immediately after the collapse of the Granite Mountain shaft, began with a stark statement. "Dear Pet—This may be the last message you will get from me." Moore gave a quick description of the events that led him and the others to the bulkhead. "The gas broke about 11:15 P.M. I tried to get all the men out, but the smoke was too strong. I got some of the boys with me in a drift and put in a bulkhead."

Moore's first note also contained his will. He discussed the disposition of a $5,000 life insurance policy, a $3,000 savings account, and other moneys—assets far above the typical miner.[27] "If anything happens to me," wrote Moore, "you had better sell the house . . . and go to California and live."[28] J.D. Moore's mother, two brothers, and a sister lived in San Jose.

In the last portion of his first note, the hard-edged Moore gave a brief glimpse of a softer side. "You will know your Jim died like a man and his last thought was for his wife that I love better than anyone on earth . . . We will meet again. Tell mother and the boys [Moore's brothers] goodbye. With love to my pet and may God take care of

you. Your loving Jim."[29] Reflecting, seemingly, the official status that
Moore intended to bestow on his will, he also signed the document
"James D. Moore."

About an hour after the first note, at 5:00 A.M. on Saturday morning, Moore wrote his next installment. "Dear Pet: Well, we are all waiting for the end . . . I guess it won't be long . . . We take turns rapping
on the pipe, so if the rescue crew is around they will hear us." Like
Duggan, Moore put his ultimate trust in a higher power. "Well, my
dear wife, try not to worry," he wrote. "I know you will, but trust in
God, everything will come out all right."

As Moore wrote—by the light of the single candle—his men took
note of his activity. Some worked on their own messages. Others, perhaps, had no paper, or could not write. Clarence Marthey approached
Moore and asked the older man to write a few words on his behalf.
Moore obliged him, adding to his own note of 5:00 A.M.: "There is a
young fellow here, Clarence Marthy. He has a wife and two kiddies.
Tell her we done the best we could, but the cards were against us."

At 7:00 A.M. Saturday, after only three hours behind the Moore
bulkhead, the compressed air hose failed completely. Moore
wrote a brief third note. "All alive, but air getting bad—Moore."

On the same page—but below his signature—Moore wrote, "One
small piece of candle left. Think it is all off."

Sixteen

•

"In the Dark"

9 A.M. In the dark.

—J.D. MOORE, JUNE 9, 1917

At 9:00 on the morning of Saturday, June 9, J.D. Moore wrote a final chilling installment in his correspondence to his wife. Unlike his other notes, written on the inside pages of a logbook, the last note he scrawled on the cover. It read simply, "9 A.M. In the dark."[1]

The last bit of candle had burned out about an hour before. After that, the men managed to light a carbide lantern. But that too failed just before 9:00 A.M.[2]

The "dark" experienced by Moore and his men was far different from the state that most of us associate with the term. Half a mile below ground and then half a mile away from the shaft, the darkness behind the bulkhead was a total absence of light. No shadows. No gray-cobalt blur. No adjustment of the eyes over time. Just black.[3]

It was a terrifying condition, and the Moore party fought desperately to reverse it. Some of the men had "moisture-proof metal containers" holding matches. For an hour they made attempts to light one, but the best they achieved was a fleeting, "dull red glow."[4] The air in the cavity would not support a flame, and no member of the party was carrying a flashlight.

Rescue was more than two days away.

At the time J.D. Moore and his men lost all light, only five hours had passed since they completed the construction of their bulkhead. How had things gone so wrong, so fast?

As with the Duggan party, the failure of candles and carbide lanterns behind the Moore bulkhead was an indication of the poor quality of the air. With Moore's party, though, the situation had deteriorated much more quickly. A combination of causes seems most likely. Gas may have entered the chamber through an undetected leak, or the poison may have entered the chamber before the bulkhead was finished.[5] The open pipe behind the Moore bulkhead was another known source of gas. The men discovered it early and plugged the end, but a significant amount of gas might already have entered their cavity—or the plug may have leaked.

The most pernicious possible source of gas was the compressed air hose itself—which the men, of course, had believed was their salvation. The *Bureau of Mines Accident Report* raises the possibility that, after the cave-in of the Granite Mountain shaft, gas may have been "forced into the broken main supply pipe." If this happened, the pipe could have seeped gas into the hose—and then into Moore's chamber.[6]

The men of the Moore party contributed to the poor air quality by failing to rotate throughout the entire interior space behind the bulkhead. According to the Bureau of Mines, the air in the chamber might have been sufficient "if the air had been circulated by moving about."[7] In the Duggan party's formal rotation, the men moved from the watch position at the bulkhead to the very back of their chamber. The effect of this human movement was to mix the air, stirring up all available pockets of oxygen.

Moore too established a watch and a rotation. Martin Garrity recalled that "[a]s soon as the bulkhead was up and in good shape we arranged to take turns watching."[8] Despite Moore's tight control of the crew, however, his rotation was less formalized than the Duggan party's. Instead of moving to the back of the blind drift when their turn was complete, the men in the Moore party appear to have clustered close to the bulkhead. Once the lights went out, of course, ongoing bulkhead inspections became impossible, and it appears that the rotation lapsed completely.

If the exact cause of the bad air remains a mystery, the manifestation of its effects was brutally obvious. "We had light up to the time the air line broke," remembered Clarence Marthey, the young father of two. "But after that . . . it was a hell on earth down there." Finding good air to breathe became an increasingly urgent issue. "Moore made us all lay face downward in the mud where the air seemed better."[9]

The juncture at which the Moore party lost all light is also the juncture at which all clarity is lost in the men's account of events. "We lost all track of time," said Marthey. "About all we could do was sleep. It was so dark we could not see each other."[10] What remains is a blurry montage of anecdotes that the participants themselves seemed unable to place into context.

Moore made periodic checks of the outside air, refusing to allow the other men to engage in this hazardous task. With each check he hoped to find that the air in the drift had cleared sufficiently to allow escape. It never did. Inside the bulkhead chamber, meanwhile, the air continued to worsen.[11]

Moore worked hard to keep up his men's spirits, offering jokes, encouragement, and assurances that they would make it out okay. For some, though, it wasn't enough. On the night of Saturday, June 9, a man described by Marthey as a "young fellow, who had more strength than the rest of us" wanted to breach the bulkhead. Moore stoutly refused, and the man backed down.[12]

There would be more calls to tear down the wall, though, and by an increasing number of the men. According to Clarence Marthey's account—later partially discredited—one miner was so intent on getting out that he struck Moore on the head with a steel bar. Though it does not appear that there were any physical confrontations, it is clear that there were sharp arguments about the best course to follow. Ultimately, Moore's insistence on staying behind the bulkhead would prevail. As with the Duggan bulkhead, it seems highly likely that a premature breach of Moore's wall would have killed the men behind it.[13]

Not that it was much better inside. Clarence Marthey recalled that by Sunday night—after more than forty hours behind the bulkhead—the men were in poor condition, "moaning, groaning, cursing and praying." There was still water, though according to Marthey it had begun "to get hot and smell bad." Many of the men were not strong enough to drink. The others used their hands to feel for the water in the dark, then "moistened their lips and poured [water] in their throats." Hunger, at least, was not an issue. This was not because the men had food—they didn't—but rather because "we were all too sick to take nourishment if it was there."[14]

The two palliatives for Moore and his men were sleep and unconsciousness. One or both explain why the miners did not respond to helmet man Tom La Martine's rapping at the bulkhead the first time, on Sunday afternoon. As La Martine banged on the timber, the men of the Moore party lay only a few feet away. The failure of the men to wake up—or revive—would ultimately exact a heavy toll.

"The air made us stupid," was how Marthey described it. "And we all went to sleep." The *Bureau of Mines Accident Report* speculates that the men of the Moore party "were probably in a semicomatose condition for at least twenty-four hours previous to being rescued."[15]

Stanley Lazykeics, the Eastern European miner who sacrificed all of his clothes, gave a succinct summary of his fifty-five hours behind the bulkhead. "Dark as hell down there . . . hot as hell . . . and hell all around."[16]

O n the surface, meanwhile, it was snowing. Sunday, June 10, 1917, a freak blizzard hit Butte. "Instead of clear skies which might have had a cheering effect on those watching and waiting for the forms of their loved ones to be brought to the surface, a blinding snowstorm added to the depressing effect of the catastrophe and the wind chilled the watchers on the hill above the mine." The blowing snow mixed strangely with the thick smoke and gas that continued to pour from the Granite Mountain shaft.[17]

The discovery Sunday afternoon of survivors from the Duggan bulkhead gave Butte a much-needed infusion of good news. For the families of the survivors, of course, the result was immeasurable joy. But even for the families of the scores of miners still missing, there was now at least some basis for hope. Against this hope, however, marched the continuing and brutal onslaught of events—the least of which was the weather.

By Sunday morning there were already sixty bodies at Butte's undertaking establishments, forty-seven of which had been identified. Throughout the day, the helmet men continued to carry a steady stream of corpses to the surface. The warm, humid conditions in the mine had caused the bodies to deteriorate rapidly, and the stench was horrid. Rescue workers tied on camphor nosebags to ward off the smell. Both the number of the bodies—and their increasingly advanced state of deterioration—forced public officials to intervene. A meeting at the Speculator was held, with participants including Mayor Maloney, Coroner Lane, and various city and county health officials.[18]

The mayor and the coroner fell quickly into a heated turf battle over the disposition of the bodies. Mayor Maloney, concerned about the potential for the spread of diseases such as cholera, insisted that the dead "be buried at once as a protection of public health." Coroner Lane, on the other hand, "was determined that the bodies should remain until identified." The two men deadlocked and neither appeared willing to budge.[19]

When Mayor Maloney's argument did not prevail, he deployed a force more powerful than any debating point—his police. He called in a squad of uniformed officers to the mine, after which "the matter finally was compromised by Coroner Lane permitting the burial of four bodies which were terribly decomposed."[20] By nightfall, ten workers at the Mountain View Cemetery would be busy digging seventy-five new graves intended for unidentified victims.[21]

The mayor's concerns about health issues—and the certain knowledge that scores more bodies were coming—led to a new process for handling the dead. Instead of dispersing them somewhat randomly to the city's many mortuaries, a central morgue was established at the Speculator. A large barnlike structure would be quickly converted to this purpose.[22]

The sudden large number of dead also overwhelmed the supply of caskets. Al Hooper, ninety-four, may be the only person alive today with personal remembrances of the disaster. He was six at the time of the fire. One of his most distinct memories is that his father, a carpenter, was pressed into service to build coffins as replacements for the diminished supply.[23]

As word of the Duggan party survivors spread through town, a new wave of family members rushed to the mine, gathering in throngs outside the gate. One of the men who was probably among them was

Ed Garrity, brother of miner Martin Garrity. For two days, Ed had been making vigilant rounds through Butte's funeral parlors. If Ed Garrity was indeed at the Speculator on Sunday afternoon, he suffered the disappointment of failing to find his brother among the survivors of the Duggan bulkhead. What he could not know, of course, was that Martin *was* still alive—holed up in a second bulkhead with J.D. Moore.

Yet as Ed Garrity continued his search on Sunday, he came across a body that he was certain was that of his brother. Devastated, he made a formal identification. On Monday morning, Martin Garrity would be listed in newspapers as one of the "identified dead." Both Ed and Martin resided at a boardinghouse called the Pacific Hotel. Martin's gregarious personality made him extremely popular with his housemates, and not even the stern photos of the era could succeed in squelching Martin's jovial smile. Ed's fellow residents joined him in mourning his brother's death.

Like Ed Garrity, other relatives of the men in the Moore party were making their own desperate inquiries. J.D. Moore's mother sent an urgent telegraph from San Jose, California, addressed to the "SUPT SPECULATOR MINE." It read simply, "IS JAMES D MOORE ALL RIGHT ANSWER IMMEDIATELY MY EXPENSE."[24]

As James D. Moore felt his life slipping away, he worried about his will. Moore—a careful planner with a life insurance policy and a substantial savings account—was concerned that the failure to have his will properly "witnessed" could create legal problems for his family. Sometime during those hazy final hours, Moore asked his men to gather around him. "You will get out alive," he told them. "And I probably will not." He asked the men to serve as witnesses to his will and then recited its contents.[25]

Clarence Marthey—the young father whom Moore had mentioned in one of his notes to his wife—remembered Moore speaking to him after discussing the will. "Encourage the others," Moore directed him.

The men "were giving up hope," recalled Marthey. "We wanted to go out in the gas and die."[26]

Seventeen

·

"WE'RE DYING IN HERE"

[C]ome as quick as you can. We're dying in here.
—UNIDENTIFIED MINER BEHIND THE MOORE BULKHEAD, JUNE 11, 1917[1]

B y the time helmet man Tom La Martine realized the significance of the bulkhead he had discovered the day before, J.D. Moore and his men had been entombed for nearly fifty-five hours—almost twenty hours longer than the Duggan party. It was 9:00 A.M. on Monday, June 11, when La Martine and three other helmet men stood again in front of Moore's bulkhead.

La Martine rapped urgently on the timbers of the bulkhead. "Hello in there!"

This time, the rescuers heard the weak voice in reply. "There are ten men in here."[2]

"Are you all well in there?" asked La Martine.

"All but two."

There must have been an enormous temptation to batter down the wall, but although the air outside the bulkhead was sustaining carbide lanterns, La Martine suspected (correctly) that deadly gas was still present. The rescuers carried no extra breathing equipment. If they tore down the wall immediately, worried La Martine, the men inside would be poisoned.

"Can you hold out a little longer?" asked La Martine.

"Yes, but come as quick as you can," came the reply. "We're dying in here."

"Take five," yelled La Martine. "We'll be back."

As fast as he dared, given the restraints of his own breathing apparatus, La Martine retraced the more than half-mile between the Moore bulkhead and the Speculator shaft. At the shaft, he rang nine bells.

On the surface, Doctors P. J. McCarthey and C. B. Rodes stood waiting for the signal. Carrying blankets, stretchers, breathing equipment, and "stimulants," they quickly descended to the 2,200 Station. From there, La Martine guided them through the twisting drifts to the bulkhead.[3]

As it turned out, by the time the helmet men tore down the wall, the air inside the Moore bulkhead may actually have been *worse* than the air outside it—or at a minimum contained less oxygen. We know this because the rescuers used a carbide lantern to light their work with picks and axes. When the lantern was held up to the hole they punched in the wall, it was "extinguished by the first air coming from the interior."[4]

Behind the bulkhead, the rescuers found Moore and his men, all clustered within thirty feet of the wall. "Only one was able to talk," reported one of the rescuers. "And he was laughing and shouting, apparently delirious from the foul air."[5] Some of the men, including Moore, were unconscious.

Martin Garrity's initial memory, after fifty-five hours behind the bulkhead, was "some water in my face."[6] Helmet men carried Garrity to the shaft, and he was the first man to be hoisted up. When the cage beneath the Speculator headframe opened to reveal Garrity, someone in the crowd shouted, "It's Manus Duggan!" The crowd erupted in joy.[7]

Garrity was completely naked, though his body was "gray from the slime of the floor." His eyes were dilated from the hours in total darkness and his lips were parched. To revive the miner, rescuers administered ammonia and then coffee. Garrity would require hospitalization, though his voice recovered fully while still at the mine. He gave rescuers his name—shattering the temporary illusion that it was Manus Duggan who had been rescued. Garrity then asked one of the mine officials, "What day is it, Friday?"

"Monday," came the response.

"This has been a double shift," said Garrity. "Maybe we will get double time."[8]

Before he was transported to St. James Hospital, Martin Garrity asked the rescuers to "notify my brother at the Pacific Hotel."[9] A reporter from the *Butte Daily Post* put in a call to the Pacific, but Ed Garrity was on his way to the Speculator morgue. Though the day before Ed had identified what he believed to be the body of his brother, enough doubt persisted that he wanted to look at the new corpses brought up that day. He had not, apparently, heard about the discovery of the survivors. A friend at the Pacific tracked Ed down and delivered the shocking news that Martin was alive.

Disbelieving, Ed rushed to the hospital. "Is it really Martin? Oh, Martin, my boy, you're really saved."[10] From his hospital room, Martin Garrity held a "regular reception" with his many friends. In characteristic form, he was "able to converse with all callers."

For anyone who wasn't already visiting Martin at St. James Hospital, the proprietress of the Pacific Hotel was busy toasting Garrity's miraculous survival with drinks on the house. "I would rather spend $20 setting up the drinks to celebrate the return to life of Garrity than

$5 for flowers for him. He is a fine fellow. Everyone knows him and likes him and we are too tickled to know what to say."[11]

R escuers at the Moore bulkhead conducted triage on the men they found behind the wall. In this case, those miners in the best condition were taken to the top for first aid, while the worst wounded were treated on the spot. None of the men could walk on their own, requiring the helmet men to lug them the half mile to the shaft. From there, they could be lifted by hoist to the surface. The rescuers fitted all of the trapped miners with breathing apparatuses, though because of the men's unconscious or semiconscious states, the effectiveness of the equipment was diminished.[12]

The breathing apparatus was certainly necessary. The air at the 2,200 level, though more oxygen-rich than the air behind Moore's bulkhead, was still thick with gas as the rescuers raced to save the lives of the trapped men. The continuing danger to both the miners and the rescuers was demonstrated by the fact that no less than eight helmet men passed out while working to bring the Moore party to the surface. None of these rescuers died, though four required hospitalization.[13]

Two of the helmet men who blacked out had overstayed their supply of oxygen. Another, Louis O'Rear, lost consciousness after one of the doctors developed problems with his apparatus. "Here is mine," said O'Rear as he stepped forward to offer the doctor his breathing equipment. "I'll take a chance." O'Rear began walking toward the 2,200 Station but passed out before he reached it. His comrades carried him out, and he was fortunate enough to recover quickly.[14]

Though various accounts are conflicting, it appears that four of the ten men in the Moore party died. Two men probably died shortly before the bulkhead was torn down; the other two died shortly thereafter.[15] Tragically, it is likely that all of the men might have been saved had the bulkhead been explored a day earlier—in the immediate aftermath of La Martine's initial discovery. Of course, none of the men would have survived if La Martine had not been able, eventually, to put the pieces together.

Like Ed Garrity and the residents of the Pacific Hotel, five other sets of relatives and friends would rejoice in discovering that their loved ones had survived behind the bulkhead built by J.D. Moore.

Clarence Marthey—the young father—did not hear the rescuers, even as they carried him to the 2,200 Station. There he revived somewhat, though he was still unable to talk when he arrived at the surface. He was taken to the engine room, where timekeeper Grover McDonald identified him. Marthey was delirious during his ride to the Murray Hospital, shouting "turn on the air."[16]

Sharing a hospital room with Marthey was another Moore party survivor, Stanley Lazykeics. Like Marthey, Lazykeics was also a father—with five children under the age of six. One of Lazykeics's friends was the first to hear that the missing miner was alive, hurrying to pass the news to his wife. She rushed immediately to the hospital, leaving her five children in the care of a neighbor. A reporter visited the neighbor's house. "Papa's at the hospital," said the six-year-old daughter, who held in her arms a sibling—one of two infant twins. "A man came and said papa was not dead in the mine, but he's at the hospital."[17]

Both Marthey and Lazykeics were covered with lacerations, the result of writhing naked on the ground in search of good air. Both would recover fully.

When Ole Erickson was pulled to the surface, he was afraid of the helmet men, thought the coffee offered to him was poison, and refused to answer questions. "What's your name?" asked one of the doctors.

Erickson's response: "You won't get me to tell." Ultimately, a mine official convinced him to reveal his identity, and his family was notified of the good news.[18]

Helmet man Tom La Martine carried the gravely wounded Mike Sullivan from the mine on his back. Early reports doubted that Sullivan would survive: "He is still unconscious and it is feared that the work of the physicians cannot offset the poison that has pervaded his system and the supreme exhaustion he has suffered." It would be more than twenty-four hours before it became clear that he would live, ultimately requiring five days in the hospital.[19]

The survivors were unanimous in attributing the credit for saving their lives. "We wanted to die and Moore couldn't let us," said Stanley Lazykeics. "And now he is dead and we are safe."[20]

"It's too bad that Moore couldn't have lived," echoed Martin Garrity. "He saved us."[21]

There are several different accounts of J.D. Moore's death, though the variations are slight. According to one, he lost consciousness only "a few minutes" before the arrival of the helmet men and died "when he was lifted on the cage."[22] A second version said he "died just as he got to the fresh air current on the station of the 2,200 level."[23] Another account said that Moore was dead when the rescuers found him—set aside at first in the brutal triage of emergency response.[24] One rescuer who was there reported that "Moore's body was still warm when we found him and while the doctors worked over him for an hour, they were unable to disclose signs of life." This rescuer believed that the miner "was dead before we reached him."[25]

One of the helmet men who assisted the doctors in efforts to revive Moore was William Budelière—the North Butte shift boss and helmet man who seemed to have been everywhere. Like Tom La Martine,

Budelière was one of Moore's personal friends. Efforts to revive Moore continued on the surface, but to no avail.[26]

I n the four days following the outbreak of fire in the Granite Mountain shaft, Butte's newspapers printed numerous reports about the successive waves of friends and family members flocking to the mines, desperate for news of their loved ones. The initial wave hit after word of the fire first spread. Another came after the discovery of the men from the Duggan bulkhead, yet another after the discovery of Moore and his men. Newspaper photos showed the throngs of people behind the head-high fences.

Holding back the frantic relatives and friends required the strength of the U.S. Army, dispatched to guard the mine's perimeter. While rescuers, undertakers, and a few miners were given access to the headframe, family members were not.

There was one exception, reported by the *Anaconda Standard* under a heading "Hope Springs Eternal." According to the story, "[a] mother who had built hopes of finding her son among those alive made such an appeal that a pass was given." In the context of hundreds of other fathers and brothers, mothers and wives, it is painful to imagine the singular anguish that resulted in her admittance. More painful still was the result. "She stood beside the receiving state where the men were given treatment and cried as she looked at each man," wrote the *Standard.* "And could not find her son."[27]

A thousand miles away, another mother sent a telegraph—her second—to the North Butte Mining Company. "IF MY SON JAS D MOORE LEFT LETTERS FOR MOTHER SISTER LEE OR BEN KINDLY SEND ME COPIES OF SAME BY MAIL LET ME KNOW COST OF SAME AND WILL SEND YOU MONEY IF HE LEFT THE ABOVE LETTERS WIRE YES AT MY EXPENSE."[28]

By the time she sent the telegraph, three days after her son's death, Moore's mother appears to know that he had left letters for his wife. In one of those letters, Moore asked his wife to "[t]ell mother and the boys goodbye." Surely this word had been passed along. Still, Moore's mother obviously hoped for some final correspondence from her son, for something more. One last word to salve the wound.

But there would be none. The North Butte Mining Company telegraphed back. "THINK ALL LETTERS WERE ADDRESSED TO WIFE AM MAILING PAPER CONTAINING THEM WILL GLADLY GIVE ANY INFORMATION WE CAN."[29]

N ot long after Moore and the last of his men were brought above ground, a signal was heard from what was believed to be the 2,600 level of the Speculator. Excitement again gripped the crowd of rescuers on the surface.[30] Could more men be alive?

It was a last false note of hope.

Eighteen

•

"POINT OF ERUPTION"

The real cause of the strike was the Speculator mine disaster
on June 8. Butte, for some time, had been a volcano
on the point of eruption.

—W. J. SWINDLEHURST, COMMISSIONER,
MONTANA DEPARTMENT OF LABOR AND INDUSTRY

O n the evening of June 11, 1917—just hours after the discovery of survivors at the J.D. Moore bulkhead—miners at Butte's Elm Orlu mine walked off the job in an impromptu wildcat strike. "The men came off the hill [and] they were absolutely unorganized," said local labor leader William Dunne. It was "as near spontaneous a strike as any strike I ever saw."[1]

While the walkout may not have been planned, the imperative to strike had been building for years—and certainly since 1914—when unionism in Butte had collapsed under the combined weight of labor infighting, Company pressure, and martial law. The underlying motives of this new strike would be venomously debated in the weeks ahead, but the immediate catalyst was clear. "The real cause of the strike was the Speculator disaster of June 8," reported the head of the Montana Department of Labor and Industry. "Butte, for some time, had been a volcano on the point of eruption."[2]

The miners outlined their grievances in a handbill they called the "strike bulletin." The cause of the "terrific holocaust at the Speculator mine," claimed the bulletin, was the relentless imperative to "GET THE ROCK IN THE BOX." This was the "devout repeat" that every miner heard "from the day he starts to work until he is burned to death or dies a victim of miner's consumption."[3]

The strike bulletin invoked the First Amendment protections of free speech and the right to assemble, and repeated the battle cry of the American Revolution, "UNITED WE STAND, DIVIDED WE FALL!" The bulletin outlined three categories of demands: reestablishment of Butte as a union town; improved safety measures; and higher wages. It called for a strike until the demands were met, and it ended with the bold-lettered statement **"We MUST have a Union!"**

The next day a mass meeting of miners was held at Finlander Hall, where the men launched the new "Metal Mine Workers' Union." They also elected officers and formulated the particulars of their demands:

1. *Recognition of the Metal Mine Workers' Union by all of the Mining Companies of the Butte District in its official capacity.*

2. *The unconditional abolition of the Rustling Card system, and to reinstate all blacklisted miners.*

3. *A minimum wage of $6.00 per day for all men employed underground regardless of the price of copper.*

4. *The mines to be examined at least once each month by a Committee, half to be selected by this Union and half by the Company, the object being to avoid, as far as possible, fires and many other accidents.*

5. *That all men starting in mines shall be shown exits to all other mines so that they shall be able to escape in case of fires or all other accidents.*

6. *All members getting seven or fifteen days lay off to be given a*

> *hearing before a Committee, three to be appointed by this*
> *Union, and an equal number selected by the Company.*
> 7. *That all bulkheads must be guarded FOR THE SAFETY OF*
> *THE MINERS by having manholes built into concrete*
> *bulkheads.*[4]

The most important demands revolved around recognition of the new union by Anaconda. Butte miners, once the foundation of the vaunted Gibraltar of Labor, had been without representation since the debacle of 1914. Without a union, the miners were merely individuals, easily replaced, against a corporate monolith of demonstrated ruthlessness and power. In the context of 1917, only a union could protect the workers' interests—including wages and safety conditions.

Integrally linked to the issue of union recognition was abolition of the rustling card. In 1914, resentment over the failure of the old Butte Miners' Union to stand up against the rustling card system had contributed to its violent downfall. The strikers of 1917 complained that the rustling card was "a vogue calculated to degrade the miner to the level of common chattel," and that "under this system there is no redress—no court of arbitration or appeal for the individual so blacklisted or eliminated."[5] For Butte miners, the rustling card was a tangible symbol of their powerlessness without a union. The miners' union even drew a link between the rustling card issue and the potent political currency of Manus Duggan, claiming that the hero of the disaster had been blacklisted from the mines for three years because he was a socialist.[6]

Beyond recognition of a new union, the miners' chief complaints were poor safety conditions and low wages. It was the North Butte disaster, of course, and the visceral shock of 163 dead, that sparked the broader strike. To this day, the North Butte fire remains the worst hardrock disaster in American history.

The miners' demands for safety inspections by a joint committee including union representatives clearly reflected their bitter distrust for both Company and state officials. To the miners, the Company's attitude toward the heavy casualties seemed cavalier at best. On the day after the outbreak of the fire, the Company-owned *Butte Daily Post* editorialized that the fatalities were "praiseworthily small"—this in the same edition whose front page featured the inch-high headline: "33 KNOWN DEAD; 162 MISSING."

As for state government officials, "No one present remembered having seen the present mine inspector," complained one handbill, "except when, after an accident, he felt called upon to rush to some mine in order to whitewash the company involved."[7] Helping to vindicate the miners' cynicism was A. E. Spriggs, a member of the State Mining Board. Spriggs arrived in Butte the day after the outbreak of the fire and declared, having barely set foot on the ground, that "we feel that the North Butte Mining Company is absolutely blameless in this matter."[8] (Spriggs, it may be recalled, was the former lieutenant governor who had colluded with Copper King William A. Clark during the Senate scandal to lure the governor out of the state.)

The miners' specific demand concerning bulkheads reflected a raging controversy about the role of doorless bulkheads in the high mortality rate. In the immediate aftermath of the fire, according to District Attorney B.K. Wheeler, rumors spread through Butte that "dead miners were found piled up against bulkheads of solid cement."[9] The official *Bureau of Mines Accident Report*—a document heavily lobbied by the mining companies—would later refute that the bulkheads played a significant role in causing deaths. The report stated that only "one man was found dead at a concrete bulkhead or at a brattice of any kind erected by mining companies in the connection to other mines."[10]

At least two witnesses gave sworn testimony that directly contradicts the finding of the *Bureau of Mines Accident Report*. At a 1918 labor trial in Chicago, the North Butte disaster became an issue. A

Croation named John Musevich—who participated in the rescue—testified in federal court that he had been with a group of helmet men who found nineteen dead miners at a company-constructed bulkhead.

"Dead?" asked an attorney.

"All dead, yes . . ." answered Musevich.

". . . How far were they from the bulkhead?"

"They was right there in the bulkhead."

"But if it had been open, the men could have got through?"

"Oh, yes, if that was open the men could go through the High Ore."

"And saved their lives?"

"Sure."

"Did you notice the fingers of any of those men?" asked the attorney.

"Oh, yes, they was all wore out, working to save themselves."

"They were what?"

"They was wore out. It was terrible to see it. It was bad to look at it."[11]

At the same trial, a witness named John M. Foss gave testimony about three of the bodies he saw. "The hands of these three miners were worn down to the second knuckle on their fingers, the bone sticking out."[12]

Whatever the number of deaths they caused, no one refutes the existence of a large number of doorless bulkheads between the North Butte and the connecting Anaconda properties. Indeed the Company newspapers were filled with first-person accounts in which survivors reported the impediment to escape posed by bulkheads.[13] Nor is there any question that Montana law in 1917 made it illegal to maintain bulkheads without doors.[14] Thus it is somewhat shocking to read in the *Bureau of Mines Accident Report* that "[f]rom the viewpoint of Butte mining men and the Montana State laws, the North Butte mine was perfectly safe as regards means of escape in case of fire."[15]

In addition to recognition of the new union and improved safety measures, the miners also demanded an increase in wages: "A minimum wage of $6.00 per day for all men employed underground regardless of the price of copper." The prevailing wage at the time was $4.75 per day, a sliding scale rate that automatically decreased if the price of copper went down, and increased—at least in theory—if the price went higher.

The 1917 rate of $4.75 was greater—by 35 percent—than wages paid to miners in 1913. The war in Europe, though, had caused a dramatic rise in inflation. "Wages," claimed the miners, "have not increased nearly in proportion to the cost of living."[16]

Anaconda immediately labeled the wage demand "exorbitant."[17] By several more objective sources, however, it appears that the miners' claims about the effects of inflation were valid. According to the Montana Department of Labor and Industry, for example, "the wage earner has been receiving higher wages than in former years, but it is also shown that many of the necessities have increased out of all proportions."[18]

Between 1914 and 1918, retail food prices increased 59 percent and for "commodities, taken as a whole," the inflation rate was 100 percent. Increased wartime demand, meanwhile, had created a period of high profitability for the copper industry. "[T]he workman," stated the Department of Labor and Industry, "feels that the profits of his employer have greatly increased and that he is entitled to a proportionate advance in wages so that his earnings may have the same purchasing power as formerly."[19] Adding to pressure on Butte miners, according to a report from B.K. Wheeler to the Department of Justice, was the fact that the "cost of living in Butte has increased far in excess of the general increase which is prevalent throughout the country."[20]

Aside from the corrosive effects of inflation, miners hated the sliding scale rates. Part of this resentment, no doubt, reflected the failure of wages to slide *upward* in conjunction with several recent increases in the price of copper. By one government analysis, rising copper prices during the period of February through June of 1917 should have triggered wage increases. Wages, though, remained frozen.[21]

B eyond the merit of their specific demands, strike leaders knew that ultimate success would depend on the meticulous execution of a highly disciplined strategy. Two elements of this strategy were key. First, the new Metal Mine Workers' Union would have to walk a precarious middle ground between two feuding labor camps of the day—the American Federation of Labor (AFL) and the Industrial Workers of the World (IWW). The internal labor divisions that had helped to doom the miners in 1914 were more potent than ever by 1917. One battle for workers' loyalty would play out in Butte, but such local eruptions were part of a struggle on a national scale.

To the right stood the American Federation of Labor. Butte miners in 1917 were highly suspicious of the AFL and its various affiliates. It was, after all, disdain for this more conservative wing of the labor movement—and its failure to stand up against repeated Company abuses—that created the context for the destruction of the Miners' Union Hall in 1914 (and its aftermath, the demise of unions in Butte). Despite these deep suspicions, the new union considered an offer to affiliate with the AFL. From a purely pragmatic standpoint, the AFL was a major national power with the potential to bring national-scale labor solidarity to bear. There was also the possibility— precisely because the AFL was viewed as relatively moderate—that the Company would negotiate with it.[22]

Even as the new Metal Mine Workers' Union contemplated the best way to deal with the AFL, they faced a far more difficult challenge on their left flank from the Industrial Workers of the World, usually referred to as the "IWW" or the "Wobblies." Founded in 1905 and headquartered in Chicago, the IWW was a radical, communist-oriented labor organization. Its constitution embraced the "historical mission of the working class to do away with capitalism." For its strategy, the IWW advocated the formation of "One Big Union," which would take down the capitalist system through a massive general strike.[23]

The IWW formally arrived in Butte with the establishment of a "propaganda league" in 1912. The organization made an early mark, ironically, not as a thorn in the side of the Company, but as a thorn in the side of Butte's then-powerful socialist party. The socialists had actually managed to elect one of their own as Butte's mayor in 1911 and 1913. The IWW attacked him relentlessly as insufficiently pure.[24]

The Wobblies flourished in the labor vacuum created after 1914. As more temperate alternatives failed, the IWW emerged as a ready and organized alternative, with the miners increasingly susceptible to "the hypnotic influences of any 'ism' that is liable to develop a leadership that can weld and steer."[25] By 1917, the Butte IWW had formed its own union. The Company, of course, did not recognize it, but by one report its local membership may have numbered more than a thousand.[26] At the time of the North Butte disaster, the IWW already was actively involved in a strike that had shut down a large portion of the timber industry in the Pacific Northwest, including the two largest mills in Montana.[27]

The IWW scared not only the Anacondas of the world, but also most Montanans as well—from socialists to the extreme right wing. Today, phrases such as "the historical mission of the working class to do away with capitalism" sound naive and vaguely quaint. In 1917, however, the context was far different. The world was at war, a war the United States had now joined. The first Russian Revolution had taken place only four months earlier. Czar Nicholas II had recently abdicated the Russian throne, and the "All-Russian Congress of Soviets" was meeting even as the North Butte disaster unfolded.

The leadership of the new Metal Mine Workers' Union recognized that affiliation with the IWW would be political poison, branding their efforts as extremist and undercutting potential support from the broader community. Thus an effort was made to disavow specifically any IWW connection. For example, the distribution of IWW handbills was prohibited at union meetings.[28] Certainly the new union's demands did not approach the radical IWW agenda.

In addition to steering a delicate path between the AFL and the IWW, leaders of the new union recognized that success would require the avoidance of violence at all costs. In 1914, the dynamiting of the Union Hall had provided the pretext for martial law, and it was under the army's protective umbrella that the Company had shut down unionism in Butte. Unlike 1914, the strike effort of 1917 would seek to avoid any activity—such as marches or demonstrations—with the potential to spin out of control. One of the early handbills circulated by the new union stated the strategy succinctly, concluding with the following statement: "IN ORDER TO WIN, WE NEED ONLY KEEP QUIET, SIT STILL AND ATTEND TO ORGANIZING, AND THIS WE INTEND TO DO."[29]

Anaconda's response to the new strike was unflinching, battle-tested, and swift.

The day after the formation of the new Metal Mine Workers' Union, a Company newspaper printed a statement on the front page. "The attack upon Butte's industries is being engineered in the main by the same element which was responsible for Butte's serious trouble in 1914," read the statement. "It is well known that recently there has been a large influx into Butte of Industrial Workers of the World and other unpatriotic and seditious persons whose one aim is to paralyze our industries, and particularly those upon which the government is dependent for its arms and munitions."[30]

First and foremost, the Company strategy revolved around affixing to the union the precise label that the miners were attempting to avoid—IWW. The recent entry of the United States into the war had provided a new degree of taint to this term. Though the IWW had not yet taken a formal position, "[t]here was considerable rank-and-file sentiment in favor of militant action against any form of conscription for the 'capitalist war.'"[31] So not only was IWW synonymous with a radical economic agenda, but it now also carried the additional stain of being unpatriotic. According to District Attorney Wheeler, the IWW name "became synonymous with 'pro-German.'"[32]

Lest anyone fail to connect the dots, the Company used its newspaper monopoly to provide a steady drumbeat of conspiracy theory. Virtually all references to the new union anchored it with the IWW link. "IWW Agitators Plan to Involve City in Serious Labor Troubles," read a typical headline. "Metal Mine Workers a Branch of the IWW," read another.[33]

The striking miners, claimed the Company press, were working directly for the Germans. "Dispatches every day bring news of troubles which have been inaugurated elsewhere, trouble in mining camps in Utah, Arizona, Colorado, Michigan, Alaska and other places." The local face might be the IWW, "[b]ut everywhere there is well-grounded suspicion that all this trouble in mining camps is fomented by German agents, who are really the men higher up, acting through local agitators and labor leaders."[34]

The *Butte Miner* was blunter still. "The number of English-speaking persons engaged in the strike agitation is remarkably small and [the] fact most of them are Finnish who favor Germany because of the European political conditions apparently shows another phase of the kaiser's activities."[35]

The second prong of the Company strategy to defeat the strike was an unwavering refusal to recognize or to negotiate with the new union. A few days after forming, the union sent a representative to present the list of demands to Company officials. "He was told by representatives of the Anaconda company that they would not deal with the organization in any manner," reported the *Anaconda Standard*. It then added smugly, "Their interview was short."[36]

On behalf of all mines, Anaconda flatly denied the existence of any legitimate complaints on the part of the miners. "No grievance of the workers has been brought to the attention of the operators and we believe that none exist."[37]

Though there were several nominally independent mining companies in Butte in 1917, including the North Butte, there was no question as to who ran the show. Anaconda spoke for all mine operators, and its ties to the North Butte in particular demonstrate the depth of its influence and power. Thomas Cole, the president of the North Butte's board of directors, was a business partner with John Ryan, the chairman of Anaconda (and a former protégé of Standard Oil's ruthless Henry Rogers). Ryan sat on the North Butte's board. The North Butte, moreover, depended on Anaconda to smelt the copper it mined.[38]

While tight ranks among the mine operators were assured, the ability of Butte workers to maintain a united front was an open question. The third prong of the Company's response to the strike was therefore tried and true. In the face of the miners' attempts to create and sustain solidarity, the Company would seek to divide and conquer.

By mid-June, the battle lines were drawn.

Nineteen

·

"FOR YOU AND THE CHILD"

Madge, dear, the place is for you and the child.
—MANUS DUGGAN, JUNE 10, 1917

In the days that followed the disappearance of her husband, Madge Duggan would be forced to endure Butte's obsessed speculation about his fate. "Everywhere," wrote the *Butte Daily Post*, "the young nipper's heroism was the principal topic of discussion."[1] The town's obsession was fed by a full-fledged feeding frenzy in the local press. Of all the stories that appeared, however, none must have been more difficult for Madge to read than a haunting piece in the *Anaconda Standard* on Thursday, June 14—four days after the rescue of twenty-five other men from behind the Duggan bulkhead.

According to the *Standard*, Manus Duggan at some point actually encountered helmet men. "It is thought that he broke down under the terrible strain," wrote the paper, and that he "turned and fled." The rescuers pursued him through the drift, according to the story. "The last report of Duggan was that he was barefooted, as the prints of his bare feet were easily distinguished by the helmet men, who were not far behind, but who were unable to catch up with him."[2]

180

If the *Anaconda Standard* account is accurate, its description of Duggan's reaction would be consistent with the effects of carbon monoxide poisoning on other miners. As earlier incidents have shown, many miners were reported as delirious, combative, and paranoid. One man refused to take coffee from rescuers because he believed it to be poison.[3] Another man in the Duggan party attempted to run away from the helmet men, though in his case the rescuers were able to catch and restrain him. Certainly the sudden appearance of the rescuers—with their face masks, breathing bags, and flashlights— would be enough to unsettle a man on the brink. And Duggan—who for thirty-six hours had tested the air outside the bulkhead with his own lungs—had probably been exposed to a higher level of carbon monoxide than any of the other men in his party. As Albert Cobb stated, "Someone said he was on the Spec Station, but if he got there he must have been plumb crazy to leave it."[4]

In a July 1917 letter to the Carnegie Hero Fund Commission, North Butte General Manager Norman Braley discussed two broad theories on why Duggan left the 2,400 Station. "Duggan either had a plan in his mind of escaping by another way, or else the long confinement had unbalanced his mind so that he did not realize what he was doing."[5] It is also possible that Duggan returned to the bulkhead to check on the men who stayed back, or that he went in search of water. By one account, Duggan, while in the station, was "talking incoherently about getting water for 'his men.' "[6]

Inconsistencies in the various reports make it difficult to know exactly *where* Duggan went after leaving the 2,400 Station—and impossible to know what he was thinking. One account did place Duggan back at the bulkhead. According to this version, the "last man to see Duggan was John McGarvey," one of the three men to stay behind the wall, unable to crawl to the station. The only way that McGarvey could have been the "last man to see Duggan" is if Duggan went back.

McGarvey told a reporter that Duggan was "staggering at the time, as though he had become weakened."[7]

Whether or not Duggan actually ran away from rescuers, it does seem likely that he was no longer thinking clearly. If he had been, surely he would have waited in the Speculator Station to see if the attempt to signal the surface resulted in contact. The men would have known, after all, in a matter of minutes.

If Duggan had a plan for another way of escaping the mine—perhaps a plan that he fell back on in a state of growing incoherence—it seems certain that it involved the Rainbow mine. One of the most consistent reports about Duggan concerned his frequent discussion of the Rainbow. "I think Duggan's disappearance can be explained by his belief that rescue was at the Rainbow shaft," said Albert Cobb. "That was all he would talk about from the time he gave up hope of getting out until the bulkhead was smashed."[8] At least two of Duggan's other companions also remembered him talking about the Rainbow.[9] Though it lay more than a mile away from the bulkhead, Duggan believed it offered the best path to safety.

Beginning on the afternoon of Sunday, June 10—when rescuers first began to realize that miners from the Duggan party were still missing—helmet men conducted nonstop efforts to locate Manus and the two other men missing from the bulkhead. Armed with the information from Duggan's companions, extra crews of helmet men focused on the passageways between the bulkhead and the Rainbow.

It was not a straight-line search. Even the most direct path was more than a mile long, and the mine workings crossed three dimensions. Not only might Duggan be moving horizontally through the mine, but also vertically, up and down dozens of potential pathways. The gas situation had improved but remained uncertain, so helmet men were still encumbered with their heavy equipment—which also limited to two hours the time that any crew could spend below ground.

One crew managed to descend the Rainbow all the way to the 2,000 level before encountering deep water and thick smoke. Other crews descended the Speculator to the 2,400, then worked their way as far as they could in the direction of the Rainbow. But there was no sign of Duggan.[10]

One of the helmet men assisting in the search was William Budelière—the North Butte shift boss who had waded through water up to his armpits in the early hours after the fire. Budelière had an extra incentive as he labored to find the missing nipper: Just as he had known J. D. Moore, he also knew Manus personally.

After three days of fruitless searches, Budelière was leading a group of helmet men on the night of Thursday, June 13. They were exploring the 2,600 level—in an area between the Duggan bulkhead and the Rainbow—when they found two bodies. One corpse leaned against a ladder with his hand still resting on a rung, while the other was in a sitting position. Budelière believed that the body on the ladder was Manus Duggan. Though both of the corpses were too badly decomposed to make a positive identification, the location of the bodies reinforced Budelière's belief that he had found the missing nipper.[11]

Word that the body of Manus Duggan had been found made its way quickly to the surface, even as the helmet men labored at the arduous task of lifting the two corpses from the depths of the mine.

It seems almost certain that Madge Duggan would have received quick notification of the discovery. Her brother, Robert Brogan, made the awful late-night trip to the mine. By this time, the temporary morgue had been established at the Speculator, and it was there that Robert viewed the body. The face he examined was too badly disfigured for him to tell if it was Manus. But the pockets revealed a definitive piece of proof. Inside of one was a seventeen-jewel Swiss watch, its case engraved in German. Manus, certainly, had no such watch.[12]

Word passed quickly back to Madge. Who but another in her situation can begin to understand what she must have felt? Could there still be a sense of relief—three days beyond the rescue of her husband's compatriots? Or, knowing by now that it was practically impossible for Manus to be alive, was the failure to find his body just a cruel extension of her pain?

There must have been times when Madge Duggan felt the trauma of the North Butte disaster would never end. By the time her brother informed her that the body from the 2,600 was not Manus, it had been six days since the outbreak of the fire. Six days of searching the newspapers and the morgues. Six days of the numbing fear that comes when the consequences mean life or death. Six days of rising and falling hopes.

As the helmet men searched for her missing husband, Madge Duggan experienced an early, bitter taste of fame—twentieth-century style. The newspapers, of course, quickly lionized both the exploits of Manus Duggan and the drama of the search to find him. "Heroism of Manus Duggan Leads to Escape of All but Himself" blared the front page of the *Butte Miner*, alongside another story headlined "Make All Efforts to Find Manus Duggan."[13] The *Anaconda Standard* was more blunt: "Duggan Missing, Fear He's Dead."

Madge Duggan, who for twenty years had lived an anonymous life, found herself suddenly besieged by reporters. What was her last contact with her husband? How did she feel? One newsman pressed for a photo of Manus. The only one that Madge owned showed Manus at age twelve in an old-style swimsuit. Madge thought the image was inappropriate for publication, but the reporter persisted. What if he cropped the head and showed Manus dressed in a formal suit? Finally Madge relented.[14] The resulting photo, forever fixing the image of the (not so) young nipper, shows the face of Duggan in a suit jacket, stiff collar, and polka dot tie. Another paper ran an artist's rendition based on the same photo.

Some newspaper stories focused on Madge herself. Most outrageous was a breathless article on the front page of the June 11 *Butte Daily Post* under the headline, "Baby Is Born to Wife of Manus Duggan Since the Terrible Disaster." The story stated that "a telephone message summoning Duggan to his home was received at the mine on Friday at midnight." This would have been a few minutes after the start of the fire. The message, according to the *Post,* informed Duggan that his wife was in labor. If Manus "returns from the tomb of the dead to the living" continued the story, "he will learn, for the first time, that he is a father." It must have been too much for the *Post* to resist: "While Duggan was leading the stricken men to safety through death dealing gases of the 2,400 a child was born to him."[15]

While certainly dramatic, the story did not have the merit of being true. As Madge read it, she could pat the large swell in her belly where her and Manus's child resided still. The next day, June 12, the *Post*'s rivals had a gleeful field day, setting the record straight. "When interviewed last night," wrote the *Butte Miner,* Madge "emphatically denied the truth of the story in last night's evening paper, to the effect that 'Baby is born to wife of Manus Duggan since the terrible disaster.' " Tweaking a bit more, the *Miner* also noted that "Mrs. Duggan is indignant over the article."[16] The *Anaconda Standard* joined in the chivalrous scolding, noting delicately that Madge was "saving her strength for her *unborn* child" (emphasis added) and quoting Madge as saying, "There's no truth to the statement that I telephoned my husband to come off shift early Saturday morning."[17]

The *Standard* was not nearly so delicate (nor so concerned about

Madge's well-being) when it ran a story two days later, speculating tactlessly that Duggan's body, when ultimately found, would be too deteriorated for a proper funeral. "It is probable that the body of Duggan, like all others now being brought to surface will be in such shape that burial immediately will be necessary."[18]

Through it all, Madge could take some refuge in the comfort of her mother's home. Her patience with the newspapers, though, had worn thin. "I have nothing to say until I get either his dead body or his living body," she told the *Anaconda Standard*. "I think he is still alive. Anyway I won't give up hope."[19]

It was a hope, though, that could not be met.

Just after midnight on Thursday, June 14, six days after the start of the fire and four and a half days after the discovery of the men in the Duggan party, helmet men exploring a manway between the 2,100 and 2,200 levels came across three bodies. Two of the bodies were found lower down, but the third man had managed to climb within ten feet of the 2,100. Had he been able to keep climbing, he would have been on a path to the Rainbow.[20]

The two lower bodies were in an advanced state of decomposition and the contents of their pockets revealed no clues to their identity. The facial features of the third man were also distorted, but in his pocket were two notes. One was signed "Duggan"; the other "Manus." Almost thirty stories above the bulkhead, the helmet men had found the missing nipper.

Because of the difficult terrain where Duggan was found, it took nearly five hours to remove his body from the mine. The helmet men wrapped the body in canvas before using ropes to pull him up to a level drift. Once on the surface, the body was transferred to the public morgue, where it was placed in a casket made of copper.

Madge's brother accompanied her to the morgue, where she read the two notes found in her husband's pocket. The first piece of paper had no heading and seemed to contain two installments: One was hopeful and upbeat, while the other described rapidly deteriorating conditions.

The second note was written in those final few hours behind the bulkhead. It was Manus Duggan's will:

To my dear wife and mother:

It breaks my heart to be taken from you so suddenly and unexpectedly, but think not of me, for if death comes it will be in a sleep without suffering.

I ask forgiveness for any suffering or pain I have ever caused. Madge, dear, the place is for you and the child.

Manus

After reading the two notes, Madge took her husband home.

•

Twenty

"DUPES AND CATSPAWS"

*[L]abor agitators . . . are but dupes and catspaws of
the agents of Germany.*

—EDITORIAL, *ANACONDA STANDARD,* JUNE 21, 1917

By June 16, 1917—eight days after the outbreak of the North
Butte disaster—the "recovery" phase was over. One hundred
and sixty-one bodies had been removed from the mines, and
the mine office said the death toll "would not exceed 163."[1] Though
the final toll had been calculated, the drama stemming from the disaster
was far from over.

For its part, the new Butte Metal Mine Workers' Union was picking up steam. On June 18, the members of Butte Electrical Union No.
65 voted to join cause with the miners. (Though Anaconda had de-
recognized the large *miners'* union in 1914, it continued to conduct
business with a number of smaller, so-called craft unions, including
the electricians.) The electricians, seeking higher wages, had actually
voted to authorize a strike on June 8—hours *before* the outbreak of the
Granite Mountain fire—but they had not yet walked off the job.[2]

Now the electricians were threatening to follow through with the
strike unless their demands were met *in addition to* the "demands

recently made by the miners." When the Company refused to deal, accusing the electricians of enlisting with the "new IWW organization,"[3] the electricians walked out.

Even as Anaconda continued its efforts to portray the strike as a sinister IWW/German plot, it also attempted to downplay the impact. "[T]he so-called strike engineered by IWW and professional labor agitators has apparently proved a complete fizzle," wrote the *Butte Miner*.[4] The Company-owned newspapers put Anaconda's spin on every event. When thousands of people attended a union meeting at Columbia Gardens, the papers claimed that attendees had been "attracted to the pleasure resort by mere curiosity."[5] The *Anaconda Standard* ran an article with the headline "All Shafts of Anaconda Company Operating."[6]

In an attempt to counter the Company monopoly on newspapers, the miners circulated handbills—"strike bulletins"—providing their version of events. The primary author of the bulletins was William Dunne, an officer in the electricians' union. "The papers insisted that all the mines were working," reported one early edition, "when every one knew that statement was false."[7]

While handbills provided a rudimentary form for communicating (and shaping) the views of the strikers, labor leaders knew they needed more. In the weeks following the strike, a remarkable series of meetings took place between strike leaders and District Attorney B.K. Wheeler. The subject of the meetings: formation of a pro-labor newspaper.[8]

In the district attorney, the miners found a sympathetic ally. Wheeler was disgusted by the biased, Company-dominated press coverage (of which he was a frequent target), even going so far as to complain directly to his seniors in Washington. "I do not state this hastily," he wrote in a letter to the U.S. attorney general, "but it is a fact that the Press of Montana generally have published reports of labor conditions which are in truth unfounded."[9]

Wheeler and several other Butte attorneys would help to arrange for a $12,500 loan to launch the labor newspaper—ultimately christened the *Butte Bulletin* in homage to its *Strike Bulletin* roots. William Dunne would be its editor.[10]

Whatever Wheeler's degree of frustration with the Anaconda-dominated press, his contribution to the launch of an avowedly political newspaper was a highly questionable action for an officer of the federal judiciary. He quickly backed away from the venture, perhaps after giving more serious thought to the conflict of interest it represented. The *Butte Bulletin*, however, took hold. It was as biased toward labor as were the Company's papers toward management. Though Butte was still without an objective source for news, the advent of the *Bulletin* at least gave voice to a second perspective.

When the absence of miners from the mines could no longer be flatly denied, the Company devised a new theory. The absentees, it was claimed, could be explained by attendance at the many funerals for victims of the fire.[11]

For one day, anyway, the claim that funerals shut down the mines might have been plausible. Less than two days after the discovery of Manus Duggan's body, Madge laid her husband to rest. On June 15, at 9:00 A.M., the copper casket containing Duggan's remains was removed from the home of Madge's parents to the Sacred Heart Church. Thousands attended the service, most of whom were unable to squeeze into the church. All during the service, a crowd stood outside in quiet respect. Inside, Father O. D. Barry eulogized the fallen hero before leading the mourners in the closing hymn, "Nearer, My God to Thee."[12] Madge would later tell the story of one small comfort during the service—the distinct flutter of the unborn child inside her womb.[13]

When Madge stepped out on the street following the service, leaning on the arm of her brother, she was greeted by the large crowd of mourners. She began to sob at the sight. Her tears, according to one newspaper account, "were unreserved." A hundred of Manus's friends and family members marched in front of the casket, "followed by Mrs. Duggan, who bore up bravely."[14]

Manus Duggan, thirty, was buried at St. Patrick's Cemetery beneath a granite stone carved with a cross. His epitaph quoted from the Gospel according to John: "No greater love hath any man that he lay down his life for his friend."[15]

B y the week after Duggan's funeral, the vast majority of the dead miners had been buried. "It is probable that conditions will steadily improve this week," speculated the *Anaconda Standard,* "as most of the funerals of the victims of the Speculator mine fire have taken place."[16]

The prediction proved false. Instead of contracting, support for the strike continued to expand. The solidarity of the electricians' union brought endorsements and "sympathetic strikes" from several more unions, including the Silver Bow Trades and Labor Council, the boilermakers, and the machinists. Significantly, though, while these new unions expressed their support for the *electricians,* they did not endorse specifically the efforts of the *new miners' union.*[17]

Still, by the end of June 1917, no amount of Company spin could disguise the facts: 15,000 of an estimated 16,500 workers were on strike. Mining activity ground to a virtual halt. For the miners, it was the zenith.[18]

By this point, the strike had grown into an event of national proportion, with prominent coverage in newspapers including the *New York Times.* The North Butte disaster already had drawn the eyes of the country to Butte. Now the strike, like the sound of distant gunfire, was attracting the attention of all factions. A major battle was forming up, and contending forces began to marshal in Butte.

For the federal government, the strike had significant implications for the war effort. Butte was one of the country's major sources of copper, and increasing reliance on electricity and telephones made the red metal a vital, strategic resource. Not even bullets could be manufactured without copper, with every shell casing requiring a full ounce of brass (made from copper and tin). Washington sent in a mediator, W. H. Rodgers, from the Department of Labor. Rodgers, though himself a former miner, never gained the trust of the new miners' union. His primary mission was to return the mines to operation, and he alienated the miners by urging them to return to work *before* settling on terms.[19]

Representatives of the national unions also traveled to Butte, including senior officials from organized labor's two poles—the AFL and the IWW.[20] For its part, the AFL wanted to keep the strike from spinning out of control. The new union needed to be brought under the AFL's wing—and quickly. For the IWW, by contrast, chaos was not such a bad thing. Certainly the fiasco of 1914 had worked to the benefit of the radicals. Now, the national attention drawn to the strike was wonderful publicity—another step toward the "One Big Union" and a general strike that would bring down the capitalist system.

For its part, the Company imported troops in the form of spies and gunmen, often "detectives" hired from prominent agencies such as Pinkertons and Browns. The miners' union claimed that Anaconda brought in "over a thousand detectives," a figure that seems inflated. Still, even by more conservative estimates, it is clear that the Company imported hundreds of spies and "goon squads."[21]

With mining in Butte at a standstill, Anaconda was now fully attuned to the seriousness of the crisis it confronted. The Company redoubled its relentless efforts to break the strike at every fault line, and in an atmosphere made all the more volatile by the war, there was no shortage of fissures to exploit.

One obvious target was the alliance between the electricians and the miners. Peeling off the electricians' union—and the other unions that had followed it—would represent a major blow to the solidarity of labor. While the Company continued to refuse even to meet with the new miners' union, it entered serious negotiations with the electricians. W. H. Rodgers, the federal mediator, was a strong advocate of dealing with the unions separately. Initially, though, the electricians refused to break ranks. On July 3, they rejected Anaconda's first offer for settlement.[22]

From the outbreak of the strike, Anaconda had worked to divide the miners along ethnic lines, a fissure always ripe for exploitation. In this case, the Company trotted out the Serbs. "Serbians Announce Determination to Continue Mining Copper," crowed the *Butte Miner*. "They will not be swerved by any agitation and preachment of IWW."[23] Any Serbs supporting the strike, claimed the *Anaconda Standard*, had been brought in by the IWW "from the East."[24] Nonsense, rebutted a miners' handbill. "The Serbo-Croatians, who are always Union men, were insulted by these articles, and joined the Union faster than ever, and even expelled from their societies men who persisted in working."[25]

To counter the impression that the Irish supported the strike, the Company papers gave prominent coverage to the formation of a new organization, the "Irish-American League of Butte." Speakers at an organizational meeting were led by Mayor Maloney, who said "the Irish of Butte would again show their love for America."[26]

One of the Company's most effective tools in combating the strike was its spies. The miners claimed that Company-hired provocateurs had been behind the dynamiting of the Union Hall in 1914, "posing as union men."[27] No hard evidence has been presented that this was true. Still, most historians have little doubt that Company spies and gunmen were crawling over Butte by June 1917,[28] and union leaders believed strongly that the spies were working to incite violence. "We know that these detectives pose as Union men and that they are outspoken against the Companies, and our members know, too, what these detectives are trying to do."[29]

There is anecdotal evidence that Company spies did attempt to spur violence. At one meeting of the new miners' union, a man claiming to be IWW jumped up and urged the strikers to "blow up transformers." The next day, District Attorney Burton K. Wheeler sent out federal officers to bring him in. According to Wheeler, "I said to him, you're a detective aren't you? He hung his head and said yes." Wheeler promised to prosecute the man if he made any further calls for violence.[30]

Whether or not the spies were provocateurs, they were certainly a valuable source of information about the union's tactics and morale. The widespread presence of spies also created an atmosphere of deep suspicion among the miners, turning one against the other. Who, among the thousands present at some meetings, might be working for the Company?

Throughout the strike, Anaconda continued to hammer on the most volatile fault line of all—constantly painting the strikers as pro-IWW and therefore antiwar. "This concerted movement in all mining camps amounts in fact to a war against the United States and in many places working men are playing into the hands of the nation's enemies" read a typical editorial in the *Anaconda Standard*. "[L]abor agitators are but dupes and catspaws of the agents of Germany."[31]

Virtually every day during the early weeks of the strike, Company newspapers kept up the drumbeat of allegations. One front-page story showed a telegraph, written in Finnish, addressed to Bill Haywood, national head of the IWW. According to the translation provided by the paper, the telegraph said, "Send Italian, Austrian and English speaking organizers here to Butte. Situation is ripe."[32] Whether the document was authentic, of course, is impossible to know.

Another newspaper story stated that police had received a report that "there was stored a quantity of arms" at the Finlander boarding-house, a building "said to be a rendezvous for the IWW leaders." A dozen police surrounded the building and "a thorough search was made of the premises." The weapons cache consisted of "four guns" and "two Finnish dirks"—a rather paltry arms supply by western standards, then or now.[33]

While some of the Company's evidence may have been questionable, the continued reports, day after day, had a significant impact on public perceptions of the conflict. This is particularly true when the IWW stories are seen in the context of the news from the war. In the early summer of 1917, most of the news from Europe was bad. "Americans Meet Defeat in Fight with Hun Submarine" blared a typical headline. Nearly a hundred Londoners were killed in a German air raid, the most deadly to date. Closer to home was a report about the "first victim of the European war from Anaconda."[34]

With their periodic "strike bulletins," the miners made their best attempt to publish rebuttals. "They are mock patriots," claimed a handbill distributed on the Fourth of July, "merely interested in the making of profits out of Government contracts."[35]

What was the real impact of the IWW in the 1917 strike? The Democratic governor of Montana, Sam V. Stewart, articulated the view at one end of the spectrum: "The IWW element has been the primary and direct cause at all times of the labor disputes in Montana."[36] At the other end was a U.S. Department of Justice analyst, Ira Glasser, who in 1933–1934 wrote an extensive report on the 1917 strike at the request of President Franklin D. Roosevelt. FDR feared the outbreak of labor radicalism in response to the widespread economic hardship of the Great Depression, and asked for a study of past labor disputes. "The evidence . . ." concluded Glasser, showed that "the miners' strike was really a strike for legitimate trade union objectives, and that the role of the IWW was minor."[37]

Both Stewart and Glasser overstated their cases. As for Governor Stewart, his statement that the IWW was the "primary and direct cause" of the trouble in Butte was refuted by, among others, the head of his own Department of Labor and Industry. Under the heading "Mine Disaster Precipitates Strike," Labor Commissioner W. J. Swindlehurst wrote the following in his biennial report to the Montana legislature:

> *There were a number of contributory causes, such as the rustling*
> *card system, the mounting cost of living, and the activities of the*
> *Industrial Workers of the World, supposedly due [to] German*
> *propaganda. The real cause of the strike was the Speculator mine*
> *disaster on June 8.*[38]

The IWW was an organization founded on opportunism. It was born of anger against abusive conditions, and it did not thrive where more moderate, effective alternatives were present. In the years leading up to 1917, Butte saw both the abusive conditions—the rustling card, low wages, unsafe working conditions—*and* the lack of effective alternatives; in other words, the absence of a meaningful miners' union.

Still, the notion that the role of the IWW in the 1917 strike was "minor" is inaccurate. The IWW had a major presence in Butte and clearly sought to hijack the strike for its own purposes. The IWW did *not* succeed in setting the objectives for the strike. The agenda was not the radical IWW goal of overthrowing the capitalist system, but rather the more modest aims of traditional unionism—the right to collective bargaining, higher wages, and improved working conditions.

The IWW, however, had a significant impact on the *appearance* of the strike. The miners were in a political battle, and as always in politics—*appearances mattered*. One of the greatest challenges facing leaders of the new miners' union was to maintain discipline and control among a diverse confederation of miners with sometimes contending objectives. Anaconda stood ready to pounce upon any misstep, its newspapers magnifying all mistakes in the broader battle for public opinion. Without public support—the kind that F. Augustus Heinze had been able to generate in his battle with Standard Oil—the miners stood no chance. It was in this battle for public opinion that the IWW greatly undermined the effectiveness of the strike.

In some instances, the union leaders succeeded in reining in the radicals. For example, at early meetings of the new union, the IWW circulated flyers that read "tomorrow we will be satisfied with no less than the complete ownership of the mines, mills and smelters." The Company newspapers quoted the flyers as further proof of the true nature of the union.[39] Leaders of the new union quickly disavowed the IWW handbills. Then, in an effort to better control the message, they required that all future handbills win preapproval by a committee.[40]

B.K. Wheeler, in a report he prepared for the Justice Department in Washington, pointed out that the new union had considered the issue of IWW affiliation and explicitly voted it down. "My impression," wrote Wheeler, "is that the press of Montana is acting at the request of the employers to create an impression in the minds of the people that the IWW element is creating a lawless situation."[41]

Certainly IWW activists engaged in unauthorized activities that undermined the union effort. And certainly the Company-controlled press was ready to link any and all IWW outbursts to the union.

In important ways, though, the new Metal Mine Workers' Union was the author of its own difficulties. It ultimately failed to maintain the tight discipline that was essential in a struggle against the finely honed Company machine. For starters, many of the leaders and members of the new union were themselves either card-carrying IWW members or had been sympathetic to the IWW in the past. One of the leaders, Joe Shannon, was a prominent IWW man.[42] (Many years before the 1917 strike, Shannon had been represented by B.K. Wheeler in the young lawyer's first case in federal court. Shannon, charged with chasing strikebreakers out of town, was convicted and sentenced to jail. In Wheeler's autobiography, he remembered that when the marshal came to lead Shannon away, the convicted man "pulled a full quart of whiskey from his coat pocket and drained the contents without a stop." Said Wheeler, "I was impressed."[43]) Tom Campbell, another leader of the new union, was widely viewed as sympathetic to the IWW.[44]

In establishing a new organization, it can certainly be argued that these IWW men were turning away from the more radical agenda of their past. Even if true, however, such nuance was difficult to sustain in the explosive atmosphere of Butte in 1917.

While unauthorized IWW flyers certainly created problems, the new miners' union was also undisciplined with its *own* language. The miners' first strike bulletin, for example, employed not only the "united we stand" language of the American colonials, but also this phrase: "We have nothing to lose but our chains." It sounded an awful lot like Marx's *Manifesto of the Communist Party,* which concluded with the famous line, "The proletarians have nothing to lose but their chains." While it is possible that the miners' language was merely ill-conceived rhetoric, it muddled the message, lending credence to Anaconda's claim that the underlying aims of the union were not mainstream, but extremist.

Even if the union's own actions had been pitch perfect, multiple elements were simply beyond its control. At the end of the day, success would have required the navigation of a perilous—and perhaps impossible—political landscape. And as July 1917 unfolded, the terrain was about to become more difficult still.

Having rejected affiliation with the IWW, the question remained of whether the new miners' union would affiliate with the conservative American Federation of Labor. There was a powerful strategic case to be made for joining: The AFL, if supportive of the miners' goals, could bring the heft of a national union to bear. Also important, there were indications that Anaconda might actually deal with the miners as an AFL union, just as they were dealing with the electricians' union, an AFL affiliate.

Weighing against these potential benefits, there remained significant rank-and-file bitterness against the AFL—a bilious residue of 1914. In the current strike, the miners saw fresh reasons to question whether the AFL still wore the "copper collar." Immediately after the formation of the new miners' union, the head of the Montana Federation of Labor (an AFL affiliate) reportedly announced that he would not support the miners. Then Samuel Gompers, the AFL's national leader, ordered the electricians to delay their strike "until arbitration had been tried." It was an edict that the electricians ignored.[45]

Despite their considerable misgivings, the miners scheduled a vote on whether to affiliate. William Dunne, a leader of the electricians' union and editor of the *Strike Bulletin,* told the miners that he knew joining the AFL would be a "hard pill." On balance, though, he argued that the benefits justified the costs.[46]

On July 10, the day before the balloting, it appeared that the vote to affiliate would pass. Then came a crushing blow. The local AFL representative suddenly stipulated that the miners could join only if they would return to work immediately—the differences with Anaconda to be negotiated later. Furthermore, the new miners' *union* would not be accepted as an affiliate. Rather, the miners would have to join the AFL as individual members.[47]

"They have double-crossed us," vented a particularly bitter *Strike Bulletin*. "We are told that our organization is illegitimate, a veritable outlaw among respectable unions, and that unless we are baptized anew, in short unless we join the AF of L we will receive no help, in fact, will be met with powerful opposition."[48] On July 11, the miners rejected AFL affiliation by a vote of 4,528 to 437—a margin of greater than ten to one.[49]

Brutal as it was, the AFL sucker punch was but the first of two blows that would land on the miners in the span of one critical week. Two days after the AFL vote, the electricians' union settled their strike against Anaconda. The Company gave the electricians everything they wanted with one exception: It would not accept the electricians' demand that the *miners' demands* be met.[50] For Anaconda, it was a "strategic surrender." By giving a little, the Company hoped to drive another deep wedge through the labor coalition.[51]

For the electricians, who had voted to strike before the fire even began, it was an offer they apparently could not refuse. Aside from receiving *all* of their original demands, an additional factor may have played into the electricians' decision. The miners charged that national AFL officials had threatened to revoke the electricians' charter if they did not accept the Company's offer for settlement.[52]

For the miners, the binds of solidarity were unraveling in their hands. Three days after the electricians' settlement, most of the other craft unions—which had joined the strike in support of the electricians, not the miners—followed the electricians back to work.[53] The miners, for all practical purposes, now stood alone.

A t the same time that the Butte electricians were coming to settlement with the Company, the general executive board of the national IWW was conducting an emergency session in Chicago. The issue that prompted the emergency meeting was the draft, or more specifically, the official position that the IWW should adopt regarding the draft. Though many rank-and-file IWW members wanted militant action against conscription, some board members were unsure. "If we oppose the draft," warned one, "they'll run us out of business."[54]

The board member with the most strident antidraft viewpoint was a man named Frank Little, the executive chairman. "They'll run us out of business anyhow," argued Little. "Better to go out in a blaze of glory than to give in. Either we're for this capitalistic slaughterfest, or we're against it. I'm ready to face a firing squad rather than compromise."

The meeting would end with no formal decision on the draft issue, and members of the IWW executive committee dispersed. Frank Little's next stop would be Butte, Montana.

I n the three weeks that followed her husband's funeral, Madge Duggan experienced a new fear—more profound, perhaps, than anything she had felt before. Since the flutter during the funeral service, Madge had felt no movement from her unborn child. She visited doctors, who told her the baby might be dead. Her father, though, was reassuring. "When the apple's ripe, it'll fall."[55]

On July 7, 1917, three weeks after the death of her husband, Madge Duggan gave birth to a healthy baby girl. She named their daughter Manus.

Twenty-one

•

"OTHERS TAKE NOTICE"

Others Take Notice! First and Last Warning. 3-7-77
—NOTE PINNED ON FRANK LITTLE, AUGUST 1, 1917

On July 18, 1917, Frank Little, the thirty-eight-year-old executive chairman of the Industrial Workers of the World, hobbled off the train in Butte. He was five foot ten and skinny, supported by two crutches to compensate for a broken foot, encased in a plaster cast. Little referred to himself proudly as a "half-breed" Indian. He wore a trademark Stetson hat, tilted jauntily to one side. He was blind in one eye. "Bruisers didn't like to fool around with Frank," wrote one of his admirers. "Perhaps they were afraid of the deadly untamed look in that lone eye when his will was crossed."[1]

Frank Little was an uncompromising revolutionary, and even many of his ideological compatriots didn't like him. William Dunne, editor of the *Strike Bulletin,* considered Little "illiterate, embittered, and badly informed on labor problems." Little, said Dunne, was "not the type of man that I admired in the first place."[2]

Little boasted a long rap sheet of agitation and arrests. In the months before arriving in Butte he had been a leader of an IWW strike in Bisbee, Arizona, which had culminated with a local vigilante group rounding up hundreds of IWW members and marching them at gunpoint into the desert. Over the years, Little had been jailed in connection with strikes and demonstrations throughout the West and Midwest, including Missoula, Fresno, Spokane, and Peoria. In Butte, Little hoped to convert the striking miners to the IWW.[3]

In the summer of 1917, Frank Little was the last thing Butte needed. He was, without apology, gasoline on fire.

The day after his arrival, Little addressed an open meeting of the miners' union at a local baseball field. By one estimate, 6,000 people gathered to hear him speak. In a speech bursting with vitriol and provocation, Little talked about a "worldwide revolution." "We have no interest in the war," Little told the crowd, including many members of the press. "Our interest is solely with the working class. We do not care what the nations, America, England, Germany or Russia, do."

Little recounted his recent confrontation with the governor of Arizona. "Governor, I don't give a damn what your country is fighting for—I am fighting for the solidarity of labor." As for labor, Little vowed that "we will make it so damned hot for the government that they won't be able to send any man to France." American soldiers, said Little, were "uniformed scabs" and "thugs in U.S. uniforms."[4]

In the days that followed, the new miners' union invited Little to address them in closed meetings. It was an action that further muddied the underlying sympathies of the union, particularly because Little was the only outside labor leader who was offered the opportunity to speak. He urged the miners to fight harder. "Use any means necessary," he said. "It don't make any difference what those means are—but use them to win your strike." At another closed meeting, Little was more explicit: "You fellows are conducting a peaceful strike! Great God! What would Uncle Sam say to the soldiers he is sending to meet the German Army if they laid down their arms and said we are conducting a peaceful war?"[5]

On July 27, Little again spoke to a large public audience, his rhetoric ratcheting ever higher. In one part of his address he condemned President Woodrow Wilson: "Two years ago every house in the country had Wilson's picture and the words 'He kept us out of war.' But when he got into the war he told the people to shut their mouths." Little railed against the Constitution. "Look the city daddies in the face and tell them to go to hell," he urged the crowd. "Also their city ordinances and laws. The ordinance is only a piece of paper which can be torn up, and the same can be said of the Constitution of the United States."[6]

The Company newspapers had a field day, with Frank Little seeming to validate single-handedly their worst assertions about the strike. Little, said the *Butte Miner,* had "worked himself into a maniacal fury as he denounced the capitalists of every class and nationality." He had "practically threatened the United States Government with revolution."[7]

Pressure on District Attorney Burton K. Wheeler continued to mount, with the press demanding that he prosecute Little for treason and sedition. "The federal authorities not only in Montana but throughout the west, seem to be very lax in their duty," opined the *Butte Miner,* "when they allow such treasonable and incendiary agitators to travel at will around the country spreading the doctrine of hatred of this nation and its institutions." Senior Anaconda officials also met with Wheeler to convey directly their concerns about the IWW chairman. Wheeler, though, would not be railroaded. He began an investigation of Little, but insisted that the outcome would be dictated by the rule of law.[8]

For its part, the miners' union continued to send mixed messages. Many labor leaders recognized that Frank Little was undermining their efforts to achieve traditional union goals. Some even embraced a conspiracy theory—that Anaconda had imported Little into Butte in order to discredit and ultimately destroy the strike. Some union leaders apparently urged Little to leave town.[9]

Even before Frank Little came to Butte, however, the miners' frustration was beginning to show at the seams. In the aftermath of the AFL conflict and the electricians' settlement, stridency was growing in the unions' own rhetoric. "Although many people do not realize the fact," said the July 16 edition of the *Strike Bulletin,* "this nation is in the preliminary throes of a vast revolution."[10]

On July 31, Burton K. Wheeler paid a visit to Lewis Orvis Evans, general counsel of the Anaconda Copper Mining Company, a skillful attorney who had cut his teeth in the Company's epic battle with Copper King Fritz Heinze.[11] By this time, Wheeler had completed his investigation of Frank Little's inflammatory remarks and had arrived at a conclusion.

There is no record of the complete conversation between Wheeler and Evans. Wheeler may have explained that he found Frank Little's remarks offensive. A few days later, Wheeler would say publicly that "[p]ersonally I think any man who talks against the government and the soldiers who will go to France should be condemned." But the *legality* of Little's remarks, Wheeler reasoned, was a separate issue. "If there had been a law to prosecute Little my office would have done so . . . My department made a thorough investigation of the case and we could not by any stretch of the imagination have indicted Little."[12]

When Evans, who had lobbied Wheeler to prosecute Little, apparently expressed disdain for the district attorney's conclusion, Wheeler challenged him directly. "Now you're a good lawyer," he said. "Here's the law." Wheeler laid down a copy of the applicable statute—the Espionage Act of 1917. "You show me under that law where I can prosecute."[13]

The relevant provision of the Espionage Act, a federal law passed hastily in the weeks after the American entry into World War I, read as follows:

> *Whoever, when the U.S. is at war, shall willfully make or convey false reports or false statements with intent to interfere with the operation or success of the military or naval forces of the U.S. or to promote the success of its enemies and whoever, when the U.S. is at war, shall willfully cause or attempt to cause insubordination, disloyalty, mutiny, or refusal to duty in the military or naval forces of the U.S. or shall willfully obstruct the recruiting or enlistment service of the U.S. to the injury of the service or of the U.S., shall be punished by a fine of not more than $10,000 or imprisonment for not more than 20 years or both.*[14]

It was hardly a model for legislative clarity. Nor, because of the law's recent vintage, was there judicial guidance on how to interpret it. Wheeler's view was that "there was not one word in it to make criminal the expression of pacifist or simple pro-German opinion."[15]

General Counsel Lewis Orvis Evans, with the statute in front of him, could point to no provision applicable in the case of Frank Little. "His only reply," according to Wheeler, "was that district attorneys everywhere else in the country seemed to be able to find ample grounds for prosecution."[16]

Within a few hours, such abstract legal analysis of Frank Little's speeches was to become moot.

A t 3:00 A.M. on August 1, 1917, a large black sedan pulled up alongside the curb in front of 316 North Wyoming Street, a boardinghouse that stood next to Finlander Hall. The engine of the car was left running as six masked men spilled onto the street. One man stayed with the car, keeping watch, while the five others ran into the boardinghouse.

Inside, the masked men hurried to room thirty, kicking down the door. It was empty, but the noise awoke the landlady, who slept in the room next door. "Who are you men?" she asked.

One of the masked men pointed a pistol at her. "We are officers and we are after Frank Little."

Obviously afraid, the landlady told the men the correct room number. "He's in room thirty-two."

The men ran down the hall and kicked in the door, finding Little inside. There was a struggle, though Little, his leg in a cast, was no match for five armed men. They pulled him through the hall clad in only his underwear.[17] The landlady heard Little say, "Wait 'til I get my hat."

"Where you're going," shot back one of the men, "you won't need a hat."[18]

On the sidewalk outside, Little again resisted, but was forced into the backseat of the waiting sedan. The car sped off, but then stopped a short distance later. The men pulled Little from the vehicle and tied him to the rear bumper, dragging him behind the car until his kneecaps were "scraped off."[19]

A few hours later, a man walking to work found the body of Frank Little, hanging by the neck from a railroad trestle on the edge of town. Pinned to his underwear was a note, scrawled in red:

OTHERS TAKE
NOTICE!
FIRST AND LAST
WARNING!
3-7-77
Ⓛ—D—C—S—S—W—T

Little's death certificate listed the cause of death as "strangulation from hanging by rope, homicidal," meaning that Little survived the torture inflicted upon him by his masked abductors—including the dragging behind the vehicle and several vicious blows to his head.[20] Whether he was still conscious when he was hanged is not known.

The lynching of Frank Little marked a murderous escalation in the chain of events catalyzed by the North Butte mining disaster. And the controversy Little generated in life would continue long after his death, with direct implications not just for the miners in Butte—but also for every citizen in the country.

For the miners, the immediate reaction to Little's murder was one of bitterness and shock. The *Strike Bulletin* devoted an entire edition to the news: "Without a word of warning, giving him less chance than would be given a dog, a man—a cripple—was hurled into eternity by a gang of the most cowardly degenerates that ever disgraced this earth,—contemptible, despicable brutes who hid their faces behind masks that no one might see them."[21]

Burton K. Wheeler would also issue an angry statement, describing the lynching of Little as a "damnable outrage, a blot on the state and county . . . Every good citizen should condemn this mob spirit as unpatriotic, lawless, and inhuman." Wheeler added the following: "If there is no law to bring [Little] into the courts to answer for his statements—and there is no law—no violence of any kind should be administered to him." Wheeler went further, making a statement that he would later come to regret. If people were dissatisfied with the current law, they should "ask Congress to pass a law that will bring men to justice who preach against the government but the law should take its course."[22]

The *Helena Independent,* edited by a zealous firebrand named Will Campbell, took a less subtle view. "Good work: Let them continue to hang every IWW in the state." In "ordinary times," reasoned Campbell, a lynching would be "regretted"—comparable to the "application of the southern rape fiend remedy." But not so during war. Campbell claimed that it was "beyond the comprehension of the average citizen why the war department has not ordered certain leaders arrested and shot."[23]

In Butte itself, the Company press was restrained—at least by comparison. The *Anaconda Standard*'s headline read "Butte's Name Tarnished by the Stain of Lynch Law." In the same edition, an editorial called Little "an anarchist, a bully and a terrorist." "Even so, the lynching was a lawless and infamous proceeding."[24]

While condemning lynch law, the *Standard* and other Company papers made clear their view of the underlying source of the problem. The lynching, said the *Standard,* "followed the failure of federal, state, county and city authorities to properly deal with the case of Little and his fellow agitators." The *Butte Miner* adopted the same line. Little's actions, it argued, "were traitorous in character and it seems reasonable to suppose that if the officials had done their duty Butte might have escaped the stigma of this act."[25]

The newspapers reminded their readers of the meaning of the numbers "3-7-77" on the note pinned to Little, though every Montanan understood full well.[26] The numbers 3-7-77 had been the calling card of vigilantes during Montana's frontier days. Sometimes they were marked on a man's door—a warning to leave the territory. Other times they were pinned on the victim of a lynching—a warning to his cohorts.

What precisely the numbers stood for is something of a mystery. Some believe they represented the dimensions of a grave—three feet by seven feet by seventy-seven inches. Whatever the specific meaning of the numbers, the basic message was clear. "In the pioneer days when law and order could not or was not maintained by the legal authorities," said the *Anaconda Standard*, "the vigilantes did that which the law could not do."[27]

The press also speculated on the meaning of the mysterious letters in the note pinned to Little's body:Ⓛ—D—C—S—S—W—T. The *L*, which was circled, apparently stood for Little, while the remaining letters corresponded to other leaders of the strike. The clear message—Little's fate might be visited upon other labor leaders.[28]

The threat was not taken lightly. "No man is safe!" exclaimed the *Strike Bulletin*. "And today those dogs . . . those cowardly, degenerate brutes, are walking the streets of Butte—having reported to the men who are behind them that they succeeded."[29]

The murderers of Frank Little were never caught, nor, assuming conspiracy, the "men who are behind them." To this day people speculate about "who killed Frank Little?"

Conspiracy theories abound, including one alleging that Little hanged himself to avoid the draft. (According to this theory, the note pinned on Little was a suicide note—and "3-7-77" was his draft number!) At least one official of the United States Department of Justice believed that Little had been murdered at the behest of Bill Haywood, the IWW leader. "He was getting jealous of him and had him hung."

Another theory had union members killing Little because they believed him to be a Company spy. Still others claimed that Little was murdered by soldiers angry at being called "uniformed scabs"; or by anti-union miners angry at Little's notorious "scab lists."[30]

But the miners and many others believed that Anaconda was responsible. "Every man, woman and child in this county knows that Company agents perpetrated this foulest of all crimes," said the *Strike Bulletin*. The *Bulletin* also claimed to possess "sufficient evidence to indicate the names of five men who took part, every one of whom is a company stool-pigeon."[31] No evidence was ever brought forward, though it is easy to imagine a Company motive. Like the dynamiting of the Union Hall in 1914, the lynching of Little was precisely the type of event that might be calculated to prompt the imposition of martial law. In the past, martial law had always worked in the interest of the Company.

The absence of hard evidence in the Little case gave way to legend over the years (which is not to say, necessarily, untruth). In his classic 1943 history, *Montana: High, Wide, and Handsome,* liberal activist Joseph Kinsey Howard wrote that "none of [Little's] attackers was ever publicly named or tried, but there are those in Butte who will name them, and who will tell of the amazing retributive fate which has caught up with most of the murderers—violent or horrible death in one form or another."[32]

One of the men to whom Kinsey was likely referring was Ed Morrissey. Known as a violent drunkard, Morrissey had been fired from the position of chief detective for the Butte Police Department. He was widely suspected—but never charged—in the brutal murder of his wife, who had been beaten to death. Morrissey reportedly worked as a hired gunman for Anaconda. In 1922, he was found murdered in his bed, the victim of a severe beating.[33]

Burton K. Wheeler, like the miners' union, believed that Anaconda was responsible for Little's murder. Asked about Little in a 1972 interview, Wheeler related what he had told a colleague at the Department of Justice in the summer of 1917. "I think the Company had him hung."[34]

Twenty-two

•

"Spy Fever"

But to me the most bizarre element of the war hysteria was spy fever,
which made many people completely lose their sense of justice.
—BURTON K. WHEELER

Three thousand people accompanied the body of Frank Little on the long march to his grave, with thousands more lining the streets along the way. It was the largest funeral in the history of Butte, and in the tense aftermath of the lynching, violence had been widely anticipated. Although union leaders pledged a peaceful procession and warned their members against disruptions, all available police officers were on duty. Mayor Maloney took the additional step of prohibiting demonstrations, speeches, and even the carrying of banners. Ultimately, the "funeral passed under conditions of complete quiet," and the day unfolded with "no disorder whatever."[1]

It was four miles from the funeral parlor to the Mountain View Cemetery, but no hearse was used in Little's procession. Instead, "his body was carried in relays by miners."[2] The Stars and Stripes marched at the head of the cortege. Little's casket, though, was draped with a flag more appropriate to the man who preached labor over country—a red silk banner with the inscription "a martyr to solidarity."[3]

There was unintentional irony in those words. Indeed by the day of Little's funeral, August 5, 1917, "solidarity" in Butte existed as no more than an epitaph. Rather it was Anaconda's strategy—divide and conquer—that had taken firm hold.

Before Little's murder, the miners' union had already suffered twin blows, first from the AFL (which refused to affiliate with the new union) and then from the electricians' union (which cut its own deal with the Company). Having successfully isolated the miners' union, Anaconda now set about to break up the union itself.

In the week before Frank Little's murder, the Company newspapers published a settlement offer in the form of a three-quarter-page advertisement. "To the Miners of Butte" it began. Significantly, the offer was addressed not *to the union,* which Anaconda still refused to recognize, but *to individual miners.* The new offer included a wage increase, though it was tied to a sliding scale.[4] The miners had always been suspicious of the sliding scale, but in July 1917, there was particular reason for cynicism. Under wartime price controls, the price of copper was scheduled to fall in August, September, and October. By October, the corresponding wage increase would fall to only twenty-five cents a day—far short of the $1.25 raise sought by the miners.[5]

Anaconda's offer also contained a slight modification of the rustling card system, which the miners had demanded be scrapped outright. The Company offered, in essence, to eliminate the waiting period for receiving a card. In the past, miners had to wait for several weeks while Anaconda conducted a background check. Under the new offer, a miner would be issued a card immediately upon application, *with one onerous proviso*—"unless, at that time, some reason exists why he should not receive one." Sole discretion, of course, remained with the Company, with no form of appeal.[6]

The union reaction to Anaconda's offer was caustic. "What supreme effrontery!" vented a new edition of the *Strike Bulletin.*

"What colossal sarcasm!" The *Bulletin* concluded with a confident appeal. "Let us cooperate in every way with each other to make our victory, so nearly won, a certainty!"[7]

The reality, though, was that any chance for victory was fast slipping away. The day before Little's murder, the *Strike Bulletin* acknowledged that "[s]ome men have gone back to work, no doubt." For many miners, the financial pinch of the strike had become unsustainable. "[S]ome few more will return to the mines, not because they have any confidence in any promises of the mining companies, but from economic pressure." And some, said the *Bulletin,* were merely working for a few days "in order to get a few dollars to leave town."[8]

For a short time, it appeared that the lynching of Frank Little might shift the momentum back to the striking miners. Certainly the strike leaders, including William Dunne, editor of the *Strike Bulletin,* sought to capitalize on the incident for exactly that purpose. Hope was revived briefly in late August, when Anaconda's smeltermen joined the strike. But a Company offer quickly enticed them back to work. From the standpoint of the miners, the trend would continue downward.[9]

The most far-reaching development in the wake of Little's murder was not an action by the workers of Butte or the Company—but by the federal government. Under cover of darkness on August 11, 1917, federal troops occupied Butte. The government's patience, it seemed, had run out. Though Butte had been mostly peaceful since the lynching, Washington was apparently unwilling to tolerate the ongoing risk to wartime copper production. Two companies of federal soldiers took up positions on the main roads leading to Butte's mines.[10]

"It Isn't Martial Law" declared an *Anaconda Standard* headline the next morning. The troops would have the power to make arrests— "because the nation is at war"—but all prosecutions would take place in civilian courts.[11] (This was a distinction from the occupation of

1914, when Butte civilians were tried in martial courts.) As for the precise purpose of the occupation, a War Department statement declared that "all miners desiring to work would be protected." The *Helena Independent* reported that "[t]he need for a greater production of copper is said to be the reason which actuated the war department in taking a hand in Butte's labor disputes."[12]

Federal troops would remain in Butte for four years, long after the end of hostilities in World War I. A 1930s report by the U.S. Department of Justice would cite the 1917 occupation of Butte as a "good example of how the 'troop-habit' was developing under emergency conditions."[13] More recently, a 1997 study sponsored by the U.S. Army's Center of Military History concluded that the United States' 1917 declaration of war against Germany would initiate "a period of unprecedented federal military intervention in domestic disorders." Between 1917 and 1918, federal troops were called out in connection with strikes in Washington, Oregon, California, Idaho, Montana, Arizona, New Mexico, Texas, Missouri, Louisiana, Tennessee, and Georgia.[14]

It is not accurate to say that the occupation of Butte by federal troops signaled the death knell of the miners' strike. The strike—even before the occupation—was bleeding from mortal wounds. Indeed, it was checkmated almost before it began. Success would have required a unity of purpose and message that was probably impossible in the volatile, divisive environment of 1917.

As much as any other factor, the miners' strike died at the hand of union factionalism. To their credit, leaders of the new miners' union recognized at the beginning that it was essential to chart a centrist course. The problem they faced, though, was that the center could not be held. To the right of the miners were the AFL and the ghost of the old miners' union. Both had failed to stand up against the Com-

pany in support of traditional union goals: the right to bargain collectively, fair wages, safe working conditions. In Butte, the AFL had lost
its credibility as a protector of miners' interests.

The failure of labor's right flank opened the door to the left—radical unionism in the form of the IWW. At the end of the day, whether
or not the IWW controlled the new miners' union—it did not—was
not the point. The issue was whether the new union *could control the
IWW*. It could not. The IWW was not an organization designed for
coalition building or negotiation, but rather it pursued the express
goal of destroying the capitalist system. Agitation was its essence.
From the beginning, strike leaders faced an impossible dilemma: To
succeed against the awesome power of the Company, the miners'
union needed the solidarity of every Butte worker—and certainly
every miner. Yet in allowing the IWW to join their cause, they had
planted the germ of their eventual destruction. Like malaria, the IWW
proved impossible to shake.

For its part, Anaconda played out the strike with the precision and
skill of a concert pianist. It refused to negotiate with the miners' union
because it knew no compromise was necessary. It identified the multiple
fissures among Butte's workers and drove a wedge through every one. It
seared the new union with the IWW brand. And finally, it tapped skill-
fully into a deep reservoir of public apprehension about the war.

Through the fall of 1917, the strike staggered along. By October,
the War Industries Board reported to the U.S. secretary of labor that
Butte's mines were back to operating at 90 percent capacity. On
December 18, 1917, the remaining strikers finally succumbed to the
obvious. A vote to end the strike was virtually unanimous.[15] The
Company had won again.

The end of the miners' strike did not seal the lid on the furies unleashed by the North Butte disaster. Indeed, even as the strike was winding down in the fall of 1917, an infamous new phase was ratcheting up.

The effort to paint the miners' strike as a German plot was only one wave in a swelling tide, and reaction to the lynching of Frank Little would propel it forward. Burton K. Wheeler would swim against this tide, then nearly drown beneath it. Reflecting back, Wheeler called it "spy fever."[16]

Criticism of the strike as a German plot took root because many Montanans believed it was thoroughly possible. In the summer following America's entry into the war—the same summer as the North Butte disaster and its aftermath—B.K. Wheeler's office received "a thousand reports at the very least" of suspected German activity in Montana. Adding to the flurry was the controversial draft. "Montana," remembered Wheeler, "was going crazy with reports of slackers and rumors of spies."[17]

Many of the reports, according to Wheeler, "were based on feuds among neighbors who seized on the spy scare to try to settle old scores." Others were simple mistakes. One report, for example, described secret military drills by armed men in the basement of a Catholic church. Wheeler investigated, finding a local fraternal organization (quite unarmed) conducting an athletic program.[18]

No single person deserved greater responsibility for the spread of wartime hysteria than a Helena newspaper editor named Will Campbell—the same editor who, on the day after the lynching of Frank Little, penned the line "Good work: Let them continue to hang every IWW in the state." Campbell used the *Helena Independent* to provide a daily catalog of allegedly pro-German activities. He attacked a Helena-based German-language newspaper as spreading "the most despised language of history, a tongue which will be damned as the world spins down the corridors of time." He ran a front-page story claiming that the IWW was agitating to send the Apache Indians back on the warpath. Most fantastic of all were a series of *Helena Independent* stories suggesting that the German military had established an air base in the wilds of Montana.[19]

The front page of the September 1, 1917, edition ran the headline "Airship Seen Flying Above Helena—Have Germans Spy Post Near Here?"[20] Campbell fed the story for months. In October, the *Independent* offered a $100 reward to anyone who could identify the mystery airplane. "This mystery must be solved," said the paper, and asked its readers, "Are the Germans about to bomb the capital of Montana? . . . Do our enemies fly around over our high mountains where formerly only the shadow of the eagle swept?"[21]

Two weeks later, Campbell's *Independent* reported that "Helena fired the first shots discharged in America at an airplane." According to the story, a Helena citizen had "emptied a high-power rapid-fire gun at the raiders." Not to be outdone, Montana Governor Sam Stewart issued a statement. "Notify me at once next time and I will pursue in my auto," he said, suggesting that he would take along an "expert rifleman." "This thing must be run down." Numerous witnesses reported seeing lights in the sky, and one woman claimed to have seen the forms of two men in a "biplane on the pattern of the Wright machines."[22] Two *Helena Independent* staff members, out hunting, had earlier reported they were "certain they heard the exhaust of an airship over the top of a hill beyond them."[23]

Anecdotes about German airplanes seem amusing today, with landlocked Montana a rather unlikely staging ground for aerial invasion by Germany. (Among the logistical difficulties of such a plan, airplanes of the era had a range of less than 200 miles.) Yet months of stories—about spies, airplanes, IWW/German plotting—had a cumulative impact. For a readership already on edge, the *Independent* and its ilk helped to fan anxiety into fear, and fear into full-fledged hysteria.[24]

Still, had Will Campbell's responsibilities been limited to the editing of a single newspaper, history would likely view him as a picayune version of our modern political bloviators. Campbell's influence, though, extended far beyond his daily publication of a local paper. While the *Helena Independent* would serve as his mouthpiece, his true power emanated from a different and ultimately sinister source—a wartime creation known as the Montana Council of Defense.

On April 7, 1917, one day after the American declaration of war against Germany, President Woodrow Wilson sent a request to the governors of the forty-eight American states, asking each to form a "state council of defense" to assist the federal government in the war effort. State organizations were asked to create a network of local councils, connecting the federal government down to the grassroots level.[25]

The purpose of these state and local councils was to coordinate industry and manpower in support of the war effort. In Montana, with its significant agricultural industry, the primary emphasis was on increased food production. There were other goals too, including assistance in the conduct of the draft, the raising of money for the war effort (especially through the sale of war bonds), and publicity to promote patriotism.[26]

Governor Sam Stewart, chairman of the new Montana Council of Defense, acted quickly to appoint eight other members. Reflecting the emphasis on farm production, the bipartisan group included the state commissioner of agriculture as well as a prominent rancher. Others included the chancellor of the University of Montana, a bank president, and one woman—a leading member of the Montana Federation of Women's Clubs. The governor also appointed his close friend from Helena—Will Campbell, editor of the *Helena Independent*.[27]

In the early months of its existence, the Montana Council of Defense hewed closely to its original mandate. It launched a survey to determine the potential acreage available for planting, queried farmers as to their specific needs to increase production, and encouraged the cultivation of vegetable gardens. It promoted the sale of war bonds and organized a Council Speakers Committee to dispatch orators around the state to promote patriotic themes.

In these early tasks, the Montana Council was highly successful. The amount of acreage under agricultural production, for example, expanded by an impressive 30 percent in the first spring season alone. Also effective was the council's work in support of the sale of war bonds and troop enlistment; Montana more than doubled its allotted goal for the sale of war bonds.[28] As for soldiers, Montana would contribute more per capita to the war than any other state in the union— 25 percent higher than the next closest state. Forty thousand men—10 percent of Montana's total population—would serve in the armed forces during the war. Montana would also suffer the highest per capita casualty rate of any state, with 939 dead.[29] In addition to helping raise troops, the Council of Defense undertook important work in support of soldiers and their families, helping, for example, to ensure that new draftees did not default on their mortgages.[30]

Through the summer and fall of 1917, though, the tone of the council began to assume a different character. The *Helena Independent,* with Will Campbell as its editor, began to emerge as the council's "unofficial mouthpiece." Editorials that once focused on topics such as physical fitness ("get that paunch down"), volunteerism ("farm during your vacation"), and temperance ("booze and efficiency do not go together") began to take a sharper edge.[31]

Frustrated by reports of college professors questioning the war,

for example, one Campbell editorial bemoaned the lack of a draco-
nian deportation law. "[I]t would be an excellent thing if he or she
were escorted down to the sea shore or to the border and kicked out
of the country never to return." Campbell also warned about allies of
Germany (unnamed) in the U.S. Senate. "The loyal Americans are
almost ready to vote Wilson a dictator and hurry congress to intern
camps or old men's homes."[32]

For Campbell and his *Helena Independent*, the miners' strike in
Butte provided dramatic evidence of pro-German activity in the state.
In the weeks before Frank Little's murder, Campbell was among those
most critical of District Attorney B.K. Wheeler's failure to prosecute
more people under the Espionage Act. After the lynching, Campbell
warned that more would follow unless the government cracked down
on the IWW once and for all. "[T]ake a hand now and end the IWW
in the West," he demanded, or "there will be more night visits, more
tugs at the rope and more IWW tongues will wag for the last time
when the noose tightens about the traitors' throats."[33]

It would be overstatement to claim that the North Butte disaster
was the sole or even the predominant cause of the radicalism about to
sweep the state. Montana *before* the disaster was already a simmering
stew of economic, political, and ethnic tensions. What seems certain,
however, is that the disaster was the immediate catalyst to a series of
events—especially the miners' strike and lynching of Little—that
would in turn accelerate the tempo and sharpen the animosity in the
volatile months to follow. B.K. Wheeler would find himself in the center of these events.

Little's hanging marked an important juncture in the evolution of
the Montana Council of Defense, which began to move rapidly astray of
its original mandate. At a meeting in the wake of the Little lynching, the
council formulated a recommendation that the state legislature establish
a new 400-man police force.[34] Even with more police, though, the

council believed that the root of the problem lay at the doorstep of passive prosecutors—B.K. Wheeler emerged as the poster boy—who insisted upon "finding a certain law to cover minutely every point involved when a person is jailed."[35] If Wheeler wouldn't prosecute Frank Little and others under the current law, reasoned Campbell and the council, then Montana needed a new, less ambiguous law.[36]

For members of the Montana Council of Defense, an even greater frustration was that the council itself *had no legal standing or power*. President Wilson had encouraged the formation of state councils under formal mandate from state legislatures. But the Montana legislature had not yet been in session, so the Montana Council of Defense was vested with no legal authority or funding. It was essentially an advisory body. It could make recommendations, but it could not undertake actions with force of law.[37]

The council's sense of impotence was about to grow larger. In late 1917, a crotchety Montana rancher named Ves Hall was arrested for making "seditious remarks." The facts of the case were not in dispute. Hall, according to several witnesses, called President Wilson "a British tool, a servant of Wall Street millionaires"; claimed that the United States did not have the right to fight outside of its borders; and argued that Germany had acted within its rights in sinking the *Lusitania* (a British cargo and passenger ship carrying many American citizens at the time it was sunk).[38]

In January 1918, the Ves Hall case came before a Butte judge named George M. Bourquin—a jurist both respected and reviled. Bourquin's mantra was the oft-quoted phrase "This court may be wrong, but never in doubt." To an even greater degree than B.K. Wheeler, Bourquin believed that the Espionage Act should be construed narrowly—held strictly to the language of the statute itself.[39] Like Wheeler, Bourquin had been the target of widespread criticism both by the press and by the Montana Council of Defense.[40]

After hearing the evidence in the Ves Hall case, Judge Bourquin directed a verdict of acquittal—plucking the outcome away from the hands of the jury. Personally, said Bourquin, he found Hall's statements to be "unspeakable," but the issue under the statute was whether Ves Hall was attempting to obstruct America's armed forces. Bourquin noted that Hall's statements "were made at a Montana village of some 60 people, 60 miles from the railway, and none of the armies or navies were within hundreds of miles so far as appears." Furthermore, the statements had been made "in a hotel kitchen, some at a picnic, some on the street, some in hot and furious saloon argument." Bourquin called the notion that Hall's remarks had interfered with U.S. forces "unjustified, absurd, and without support in evidence."[41]

Reaction from the press was immediate and predictable. In the *Helena Independent,* Will Campbell slammed Bourquin for acquitting a man "who had slandered, libeled, lied about the country we profess to love." Campbell also wrote a letter to U.S. Senator Henry Myers asking if Bourquin could be removed from the state during the war. Montanans, he said, "are determined to rid the state of Wobblies, disloyalists and traitors."[42]

When piled on top of frustration with B.K. Wheeler over his handling of Frank Little, Judge Bourquin's decision in the Ves Hall case would be the last straw. A week after the decision was handed down, Governor Stewart called an Extraordinary Session of the Montana State Legislature. The intent of the session was straightforward. First, the legislature would remove all ambiguity from espionage and sedition law in Montana. And second, in order to ensure that the new laws were enforced, the Montana Council of Defense would be vested with expansive new powers. The council "should be given a legal existence," proclaimed the governor. "Not only should the State Council of Defense have a legal status but financial provision should be made for the conduct of its work."[43]

What had begun as a laudable effort to mobilize grassroots support for the war effort was about to be transformed. Whipped into a fury by the tumult of 1917—including the strike in Butte and the lynching of Frank Little—an advisory body designed primarily to boost agricultural production was about to become the most powerful law enforcement entity in the state of Montana. The result would be harassment, inquisition, and some of the most dramatic nationwide restrictions on constitutional freedom in modern American history.

Twenty-three

•

"WHAT WILL BECOME OF THEM"

What will they do now? He has a wife and two small children.
God knows what will become of them.

—*BUTTE MINER*, JUNE 10, 1917[1]

adge Duggan received a death benefit of $3,380 from the
North Butte Mining Company, an amount fixed by Montana state law. It was the equivalent of approximately three
years of a miner's pay. In honor of Manus's heroism, the North Butte
made an additional gesture—paying the cost of his headstone.[2]

Other families of dead miners were less fortunate. Funeral expenses emerged as an immediate point of contention—salt on an open
wound. The state compensation law held the mining company liable
for only $75 in funeral costs, though according to Butte undertaker
Larry Duggan (no relation to Manus), "the cheapest funeral in Butte is
$90, and then it is without any profit to an undertaker." Duggan esti-
mated the cost of a "proper" burial at $190, and expenses of more
than $200 were not considered extravagant.[3]

It might be argued that mortician Larry Duggan was far from an
objective source on the reasonableness of funeral expenses. Likewise,

families in the emotional aftermath of a loved one's death might not be expected to provide the best measure of appropriate limits. Beyond either morticians or families, however, there is other evidence that $75 was a paltry sum for burial.

One of the dead miners was a Serbian named Savo Vudovar, whose funeral expenses were paid by the Servian Bocalian Brotherhood. Though Vudovar had not been a member of the society, it covered the funeral because "we felt it a duty as a fellow countryman to extend every possible token of respect to his memory." Costs included $162.50 to the Duggan Funeral Home, $30.00 for "gloves and flowers," and $25.00 for services at the church—for a total of $217.00. On September 13, 1917, the Brotherhood sent a letter to the North Butte Mining Company acknowledging the $75 paid by the company but then outlining the actual expenses. The letter asked the company for "your aid in paying this bill to as liberal an extent as you may deem best." There is no record of the company's reply.[4]

If a wife spent as much as the Servian Bocalian Brotherhood on a funeral, there were significant consequences for the compensation due to her. For wives who held funeral expenses to less than $150, there was *a choice* between (1) a lump sum death benefit of $3,380, or (2) $4,000 paid out in $10 weekly installments. If, however, a wife paid more than $150 for her husband's funeral, the State Compensation Board deemed her incapable of managing a lump sum payout. As a consequence, she could receive the death benefit only in the form of a weekly payment.[5]

The issue of funeral expenses was relatively minor compared to other shortcomings of the state compensation law. The law itself had been passed only recently—in 1915. Anaconda had fought against a law of any sort for years before the disaster. In 1914, concerned Montanans finally forced a vote on the issue through a citizens' ballot initiative. Anaconda defeated the measure through a well-organized misinformation campaign—sending speakers into rural areas to tell farmers and ranchers, falsely, that they would be liable under the proposed law for injuries to their hired hands.[6]

One of the most significant weaknesses of the law at the time of the disaster concerned the provisions for who counted as "dependents." Compensation was due only to the families of those miners who had a *wife and/or children*. For single miners, the law required only that the company pay the $75 toward burial. But in many cases, such unmarried miners supported elderly parents. Mrs. B. Thompson was the mother from Augusta, Kansas, who had sent the six desperate letters and telegrams to the North Butte Mining Company in the aftermath of the fire. Her son, Vernon, was among the dead. In one of her letters she wrote that "my husband being an invalid I depended on Vernon for everything." In another, "He [Vernon] was the head of the family."[7] The compensation law made no allowance for such circumstances.

Also unprotected were miners who were not U.S. citizens—a substantial proportion of the Butte workforce. Even if a miner *was* a U.S. citizen, his wife and children were due no compensation if they resided outside of the United States. This, again, was a common occurrence.[8]

In the early days after the fire, the Company newspapers predicted compensation payments of $500,000 to $750,000. The reality would prove far different. By one estimate, when all statutory exceptions had been applied, only 39 of the 163 families of dead miners received compensation (beyond the $75 funeral stipend). Total compensation payments would amount to less than $150,000.[9] In 1918, to correct the shortcomings in the compensation law brought to light by the North Butte disaster, a new citizens' ballot initiative was attempted. A Company-led campaign, once more, shot it down.[10]

Though her situation was obviously wrenching, Madge Duggan could count a number of blessings. Most of her family—including her father, mother, brother, and two sisters—lived in Butte. In the weeks after Manus's death, they helped to complete construction of the house that Manus had not quite finished building, and later they all moved in. What Madge sacrificed in privacy, she recouped in family support—a fact that must have been particularly important as she struggled to raise her infant daughter.[11]

Other families of dead miners stood on far less stable ground. Under a headline "Misfortune Lot of Man from the East," the *Butte Miner* told the story of the Bennetts family. Henry Bennetts had just moved to Butte with his wife and six boys from Smokerun, Pennsylvania, where the family lost their home. Henry's new mining job, he hoped, would help him to "get on his feet again." Instead, his death left his wife to raise their boys alone. They ranged in age from three to thirteen.[12]

Some children were left orphaned by the disaster. One "notable victim" of the fire was a man named Maurice Fitzharris. Before working in the mines, Fitzharris had owned a hotel and also had served in the city government. Only three days before the fire, he left his government job and took a position at the Speculator with an understanding that he would be promoted to shift boss "as soon as he became acquainted with the workings." Fitzharris's wife had died a few years earlier, and he lived alone with three of his children—the oldest just nine years old. A fourth child, aged three, lived in Minneapolis with an aunt.[13]

J.D. Moore—whose clear thinking helped to save the lives of six men during the fire—was one of the few miners with the foresight and sophistication to purchase life insurance for his family. So careful was Moore, in fact, that he apparently made sure his fellow miners witnessed his signing of the will that he wrote while trapped behind his bulkhead. Moore's assets included a $5,000 life insurance policy, a $3,000 savings account, and a house. Combined with the death benefit under the workman's compensation law, Moore's wife would have been relatively secure.

The assets of Ernest Sullau, the man who started the fire, were probably more typical. A notice in the *Anaconda Standard* on June 16, 1917, reported that Lena Sullau, age forty-four, was seeking to clear title to her husband's estate. The estate consisted primarily of "half interest in two lots in the Grand Avenue addition . . . valued at $500." Sullau, it will be remembered, was an established miner, an assistant foreman, and had no children. For younger men with families, nest eggs and safety nets were rare.

Even the families of the most senior men would face difficulties. Con O'Neill was the foreman of the Anaconda-owned Diamond-Bell, killed after rushing to the mine against the pleadings of his wife. After his death, O'Neill's wife and children found their world changed overnight. Before the disaster they lived in a "beautiful home," a property owned by the Anaconda Copper Mining Company. Along with the house came perquisites, including a maid and chauffeur. Within days after Con's funeral, however, Anaconda asked the family to vacate the premises so that the new foreman could take up residence.[14]

The family received their lump sum death benefit, which Con's wife, Julia, used to buy a new, much smaller home for herself and her four children, none older than twelve. Con and Julia owned a small piece of property in Hamilton, Montana, to which they had hoped one day to retire. Instead, Julia rented it out, generating enough income for the family to scrape by. "She had everything one day," said one of her granddaughters, "and nothing the next."[15]

It was not until the 1950s that the family made a discovery: Anaconda owed Julia the pension that would have been Con's. There had never been any word of a pension from Anaconda. Under pressure from the family and a prominent "family friend" (possibly B.K. Wheeler), Anaconda would ultimately pay Julia a "modest pension" until her death in 1955.[16]

W hatever the level of support they received, all widows of miners faced the sudden, solitary responsibility of supporting themselves and their families. In 1917, women in Butte were involved in a wide range of income-generating activities, albeit generally low wage. The most common job outside the home was work as a domestic servant. Women with extra rooms often took on boarders, for whom they might also cook, clean, and do laundry. A typical mine widow—hardly uncommon in Butte—would likely cobble together a number of odd jobs, all of which she had to arrange around child care. One widow, who worked as a midwife, also "had a cow and she baked bread and she used to take in washing and then the children would deliver the washing. And that's how she raised her children."[17]

Of all the widows of the North Butte disaster, none followed a career path quite as remarkable as that of Madge Duggan. In an odd providence, her husband's heroic death presented opportunities that otherwise would have been closed to her. To Madge's great credit, she pursued each chance with moxie and skill.

In the summer of 1918, a year after Manus's death, local Democrats urged Madge to run for the office of public administrator. Though Montana's Jeanette Rankin had been the first woman elected to the United States Congress (in 1916), women candidates were hardly common. It had been only four years since Montanans approved a state constitutional amendment granting women the right to vote, and a national suffrage law was still two years away. At the time of the November election, Madge was barely old enough—twenty-one—to vote for herself.

When the votes were counted, though, Madge won a decisive victory, outpolling her opponent by 59 to 41 percent. Of the more than twenty races decided on election day, Madge received the highest vote count of any candidate.[18] Her opponent, a Mr. W. E. Coman, complained bitterly to the press. "[T]he Spanish influenza epidemic," he claimed, had prevented him from "campaigning to offset the effect of 'sympathy' stories."[19] Coman's snide comment, aside from the insight it revealed about his character, forewarned of the immediate challenge that Madge would face in her new job.

The winter of 1918 to 1919 witnessed the worst natural catastrophe of the twentieth century—a deadly epidemic of Spanish flu. (The origin of the disease, despite the name, was a strain of bird flu from China.) It is estimated that half of the humans on the planet became ill, and *twenty-five million* people died. In the United States, half a million died. Montana had the fourth highest infection rate of any state, and approximately one in every 100 Montanans perished from the disease. One-third of Montana's deaths occurred in Butte-Silver Bow County. Sadly, wild public celebrations of the Peace Armistice, on November 11, 1918, accelerated the spread of the disease.[20]

As public administrator, one of Madge's responsibilities was to locate the relatives and settle the estates of those who died. It was exhausting work, both emotionally and physically draining. It was also a job with considerable danger. At the height of the epidemic, a dozen or more Butte residents were dying *each day*.[21] While most residents were warned to stay home, Madge often journeyed to the homes of the dead. With Butte's transient, cosmopolitan population, locating relatives sometimes required considerable detective work. Madge must have been particularly well suited to deal with distraught families.

Certainly her own recent experiences had endowed her with the gift of empathy.

Madge won reelection in 1920 by a narrower but comfortable margin.[22] Fortunately, there would be no adventure comparable to the 1918 flu epidemic in the balance of her term. There was, however, an amusing encounter that offered an insightful glimpse of Butte-style politics. Among the election-rigging techniques for which Butte was notorious, one common practice was to vote under the names of the recently deceased. At one point during Madge's term—the precise date is not clear—she worked as an election official at one of Butte's polling stations. A man came in to vote, and when Madge asked him his name, he said, "I'm Manus Duggan."

"Oh, you are?" replied Madge, without missing a beat.

"Yes."

"Well, I'm pleased to meet you," she said. "I'm your wife."

The man, suddenly flustered, ran away.

As Madge approached the end of her second term in 1922, friends encouraged her to seek the higher office of county treasurer. They believed that the combination of Madge's job performance and minor celebrity status would make her a shoo-in. Her potential opponent—the incumbent—agreed. He approached Madge and made a direct appeal. "Don't run against me, Madge," he said. "I need the job. I've got five kids to support."

Madge decided not to run, though her professional career was not over. The more mundane responsibilities of Madge's public administrator office had included a significant amount of bookkeeping. At the time she was first elected, Madge had only an eighth-grade education, and her only professional training had been as a seamstress. To improve her skills, Madge began taking accounting courses at the Butte Business College. She began a second career as a bookkeeper—working for both Winona Oil and the Roundup Company.

Though she would never again seek public office, Madge remained politically active throughout the rest of her life. She was a strong supporter of the Democratic party in general and of B.K. Wheeler in particular. Madge campaigned for Wheeler, and "he was always up at the house."

In 1927, Madge married Tim O'Conner. She stopped working as a bookkeeper at that time, but would later open her own dressmaking shop. Madge's daughter remembered O'Conner as a loving father and husband.

Madge Brogan Dugan O'Conner died on October 5, 1967, at the age of seventy. She was buried between the two men she loved in life— Manus Duggan on one side and Tim O'Conner on the other.

Twenty-four

•

"SOME LITTLE BODY OF MEN"

How would you like to have some little body of men
get together in secret meetings and pass resolutions?
—BURTON K. WHEELER, TESTIFYING BEFORE THE
MONTANA COUNCIL OF DEFENSE[1]

O n February 14, 1918, Governor Samuel Stewart rose to address the members of the Montana State Legislature, assem-
bled by him in Extraordinary Session. He warned the
legislators of "traitors in our midst," "vipers circulating," and "poisoned tentacles." To address this danger, argued Stewart, it was necessary to enact "some new laws, some new working rules, some new
plans and specifications." Before February was over, the necessary legal changes would be in place.[2]

For Will Campbell, editor of the *Helena Independent* and member
of the Montana Council of Defense, it was a moment of supreme sat-
isfaction. His relentless crusading had borne fruit. The days of watching the government in helpless frustration were over, because the
Montana Council of Defense was about to *become the government*.

The Montana Council of Defense Act, passed in late February,
vested the council with a stunning array of new powers. It was granted
the right to "do all acts and things not inconsistent with the Constitution

or laws of the State of Montana, or of the United States, which are necessary or proper for the public safety and for the protection of life and public property, or private property." The council was given the specific power to create orders and rules that would have the same legal force as acts of the duly elected legislature. All state government offices and officials were placed at the council's disposal for "aid and assistance." Any person violating an order or rule of the council was subject to fine and imprisonment.[3]

With the mechanism for enforcement in place, all that remained was to clear up the ambiguities in the law. Two weeks before the Extraordinary Session, Will Campbell had called for passage of a state sedition act "which will get every offender behind the bars who cannot be reached through the federal courts."[4] The legislature responded with the Montana Sedition Act. Unlike the ambiguous federal Espionage Act, the new law was crystal clear:

> *Any person or persons who shall utter, print, write or publish any disloyal, profane, violent, scurrilous, contemptuous, slurring or abusive language about the form of government of the United States, or the constitution of the United States, or the soldiers or sailors of the United States, or the flag of the United States, or the uniform of the army or navy . . . or shall utter, print, write or publish any language calculated to incite or inflame resistance to any duly constituted Federal or State authority in connection with the prosecution of the War . . . shall be guilty of the crime of sedition.*

Persons violating the new statute faced penalties of up to twenty years in prison.[5]

The Montana Sedition Act had direct links to the North Butte disaster and its immediate aftermath. The language of the statute was virtually identical to a bill drafted by one of Montana's U.S. senators, Henry Myers. On August 15, 1917—two weeks after the lynching of Frank Little—Senator Myers had introduced his original version of the legislation on the floor of the U.S. Senate, explaining that he drafted it in response to the failure (of B.K. Wheeler) to prosecute Little. Senator Myers's federal version remained stuck in the Judiciary Committee, though not for long.[6]

As for Montana's new state law, the *Anaconda Standard* summarized, approvingly, the new legal regime: "There is no freedom of speech any longer for the disloyal or pro-German. A man can talk all he pleases if he talks right."[7]

Armed with the restrictive language of the Montana Sedition Act, the newly potent Montana Council of Defense was primed for action. The legislators of the Extraordinary Session, though, had a few more axes to grind. They passed into law the Criminal Syndicalism Act, effectively making the IWW an outlaw organization. So deep-seated was fear of pro-German activity that the Montana Legislature went so far as to pass a law requiring the registration of guns.[8]

Finally, though the legislators had no power to remove federal officials, they attempted to pass resolutions calling for the resignation of both Judge Bourquin and District Attorney Wheeler. The Bourquin resolution was set aside. The anti-Wheeler resolution, however, came to a vote—failing by the thin margin of 29–30.[9]

Will Campbell would write a bitter editorial about the failure to pass the anti-Wheeler resolution. After noting Wheeler's "very narrow escape," Campbell warned the DA's political patrons that "there is strong feeling in this state against the young man from Butte because of the failure of successful prosecution in the slacker and espionage cases."[10] Neither the legislature's resolution nor the Campbell editorial would write the last chapter in the editor's rivalry with Wheeler. With the new powers granted to the Montana Council of Defense,

Campbell would no longer need to content himself with the comparatively passive act of publishing his views in the newspaper. Now he could act—and act he would.

Two weeks after the conclusion of the Extraordinary Session, the Montana Council of Defense convened in the state capitol building. Their first action was to decide that all council meetings would be held in private, with outsiders participating only upon invitation. Among the other effects of this rule, Will Campbell became the only member of the press with access to all council proceedings.[11]

Over the next seven months, the council would pass seventeen "orders"—all having the force of law. Some orders, broadly framed, were drafted with specific targets in mind. Order Number One, for example, stated that no "parade, procession or other public demonstration" could be held without the "written permission of the Governor as ex-officio chairman of this Council." The order was aimed at preventing a St. Patrick's Day parade in Butte, which the council feared would turn into an antiwar demonstration.[12]

Another council order prohibited the establishment of any new newspapers in the state, and also forbid any *weekly* paper from converting to a *daily*. The order was necessary, according to the council, "because of the absolute necessity of curtailing the use of paper" during the war.[13] The actual purpose was more focused. In the last month of the miners' strike, labor leader William Dunne (with the support of B.K. Wheeler and others) had transformed the ad hoc *Strike Bulletin* into a pro-labor weekly newspaper—the *Butte Bulletin*. The *Bulletin* was a harsh critic—and the only press critic—of the Council of Defense. Though the paper began its life as a weekly, the editors had made clear their intention to commence daily publication as soon as finances would permit. It was the transformation of the *Butte Bulletin* from a weekly to a daily that the council's order was designed to block. Dunne would ignore the order, an action that brought him into direct conflict with the council.[14]

Newspapers were not the only publications that the Council of Defense attempted to censure. The council undertook a spate of book banning after a Helena attorney named John Brown discovered an offensive passage in one of his children's schoolbooks. The passage, from a text entitled *The Ancient World,* claimed that "[t]he great contributions to civilization in the West were Roman and Teutonic." *The Ancient World,* Brown discovered, was used in forty-two Montana high schools. He quickly brought the text to the attention of the council. Appreciative and concerned, the council asked Brown to conduct a survey of other schoolbooks.[15]

Brown reported back with a list of nine more books, all of which the council promptly banned in Order Number Three. The banned books included such titles as *The German Song Book, Writing and Speaking German,* and *First German Reader.* One enthusiastic woman wrote to the council about her community's efforts to comply with the order. "A few days ago we burned all our West's *Ancient Worlds.*" She also reported on an additional step: "We also spell germany without a capital letter."[16]

Order Number Three was not limited to banning books about Germany—it also banned the German language. Will Campbell had long been concerned about the use of German, particularly in schools. In a 1917 editorial encouraging students to study Spanish instead of German, he argued that "there will always be Germans to translate German books to sell to those who do not read German."[17] Order Number Three stated that "use of the German language in public and private schools and in the pulpits of the state . . . is hereby forbidden."

The ban on German prompted a significant amount of protest,

especially by members of the clergy. Lutherans, Congregationalists, and Mennonites were among those religions whose members included large numbers of German immigrants. Letters poured into the council. One pastor noted that many of his German members understood no English beyond the words "God, Jesus, and Amen." He even limited his request to asking for permission to conduct just the communion service in German, arguing that it was "sinful to partake in the Lord's supper without understanding."[18] His request, though, was denied.

A Lutheran minister began his appeal to the council by establishing his congregation's patriotic credentials. He noted that they had pledged "to buy $1,385 worth of War Stamps, a large sum for a congregation of a little more than a hundred souls." The minister even attached a newspaper article showing that the community's school had been the first in the country in which 100 percent of the students had purchased War Stamps. To no avail.

Though clearly made uncomfortable by the notion of impinging upon religion, the council was uncompromising. One council-produced pamphlet outlined the bottom line: "We are today either loyal citizens of this our native or adopted land, or else we are traitors. The neutral or 'half baked' citizen, in time of war is an impossible conception."[19]

Of all the orders promulgated by the Montana Council of Defense, none was more far-reaching than Order Number Seven, adopted on May 28, 1918. One power that the state legislature had not explicitly granted to the council was the ability to undertake investigations. In Order Number Seven, therefore, the council gave itself the right to conduct "hearings and investigations in all matters pertaining to the public safety and the protection of life and property."

Order Number Seven also provided for the power to compel a witness's attendance through subpoena, and to force the "production of papers, books, accounts, documents and testimony." An accompanying order on procedures allowed the council, after hearing the evidence, to "take such action and make such reports or recommendations as the said council shall deem necessary."[20]

Council members would move quickly to make full use of these new powers they had granted to themselves. Many historians call the period to follow "the inquisition."[21]

A s the Montana Council of Defense contemplated a range of possibilities for their new investigatory power, no target stood out more provocatively than the federal district attorney—Burton K. Wheeler. Council member Will Campbell, in particular, could scarcely contain his smoldering grudge. Campbell had been outraged when the Extraordinary Session of the legislature had failed to pass a resolution condemning Wheeler. Now, though, Campbell and the council could use their own powers to go after the man they saw as insufficiently vigilant in ferreting out slackers and spies.

Wheeler's job status put him in a particularly vulnerable position, because his four-year term as Montana's federal district attorney had actually expired a few months earlier, in November 1917.[22] The authority to reappoint him lay with the White House. As a matter of tradition, however, the White House tried to follow the recommendation of senators from the state in question. This tradition had particular resonance, because President Wilson and Montana's two senators were all Democrats.

One of Montana's senators—Thomas Walsh—was Wheeler's political patron. After Wheeler supported Walsh in the 1910 election (against the explicit warning of the Company), Walsh rewarded him with the district attorney position. Walsh in 1918 advocated Wheeler's reappointment to the job, though the controversy surrounding the young DA was beginning to emerge as a political liability. Montana's other senator, Henry Myers, was an outspoken Wheeler opponent. The White House, apparently unwilling to offend either Myers or Walsh, left Wheeler in place until the two senators could come together on a recommendation.[23]

Senator Myers helped to draw the Montana Council of Defense into the controversy (though it would doubtless have found the way on its own) by sending a letter that asked the council's opinion on the Wheeler reappointment. The council's viewpoint, of course, was already well known. In fact, the council was already busy with other pretexts for investigating Wheeler—particularly his supposed lack of vigilance in investigating spy suspects in Butte.[24]

The council's predisposition toward Wheeler was on full display in an incident the morning before Wheeler first testified. A joint session of state and county defense councils adopted a resolution stating that Wheeler's reappointment would be "inimical and injurious to the best interests of this state and the peace of its people." During a recess, someone pointed out that it would probably have been better to pass the resolution *after* Wheeler had been heard—and so it was rescinded (if only temporarily) in the afternoon.[25]

Wheeler would testify three times before the council. The *Helena Independent,* writing about the confrontation with Wheeler, called it "the greatest show-down ever held."[26] Indeed, the theater of the hearings was more important than the substance. The council's intention was to embarrass Wheeler and kill his chances for reappointment. Wheeler, meanwhile, embraced the opportunity to demonstrate the dubiousness of the council enterprise.

In contrast to the frightened, deferential attitude of most witnesses before the council, Wheeler went on the immediate offensive. "I have been told," said Wheeler, "that the purpose of my coming here . . . was for the purpose of trying me, so to speak." Wheeler told shocked council members that his interpretation of the law was that their appropriate role was to assist *him* and other government officials. "I personally doubt whether or not the committee has any right to issue subpoenas and compel attendance of witnesses."[27]

The council pressed Wheeler hard on the numerous controversies of the past year. Will Campbell stated his opinion of the strike in Butte. "[O]ur understanding of it always has been that the Metal Mine Workers' Union had come under IWW control. You don't consider that it had?"

"I don't really believe that," answered Wheeler. He quickly established his credentials as an opponent of the IWW. "I think that a great many of the IWWs are just a lowdown trash of the earth—there is no getting away from it." But as for the cause of the strike, "if [Anaconda] had raised the wages fifty cents a day before they went on the strike, and it had not been for this disaster, there never would have been any strike in Butte."[28]

Council members did not limit their inquiry to Wheeler's professional conduct, asking numerous questions about such private matters as his religious affiliation, friends, and personal finances.[29] In one incredible exchange, Campbell asked Wheeler a series of questions about his net worth, the number of War Bonds he had purchased, and his failure to give more patriotic speeches.

Wheeler said he was "only too glad and willing to speak upon the Liberty loan, or speak in favor of the Red Cross, also to speak in favor of any other war activity." As for War Bonds, Wheeler guessed he had purchased between $500 and $750. He balked at giving a statement about his net worth. "I don't care to go into my private worth for publication over the state."

At several points, Wheeler took his own shots at Campbell. In a discussion of the public's "verdict" on the district attorney's job performance, for example, Wheeler criticized the press for its "agitation" against him and his office. Many stories, Wheeler claimed, had been based upon "unfounded and untruthful statements." In another pointed exchange, Wheeler accused Campbell's *Helena Independent* of being "absolutely subsidized and subservient to the mining interests of the state."[30]

Anaconda's close connections to the Council of Defense were on naked display at several junctures throughout the body's existence, but no time more so than during the council's consideration of Wheeler. Prior to the hearings, the council sent a letter to Anaconda asking for the Company's views on Wheeler's reappointment (a purely rhetorical inquiry). During the hearings, witnesses from Anaconda provided direct testimony against Wheeler.[31] Most incredible, perhaps, was the role given to Lewis Orvis Evans—Anaconda's general counsel and the same man with whom Wheeler had sparred in Butte over the potential prosecution of Frank Little.[32]

The council's official legal advisor was Sam Ford—the attorney general of Montana. Ford, though, had begun to give the council members advice that they did not want to hear. In his place, the council relied increasingly on outside, unofficial advisors—especially Lewis Orvis Evans. Evans made his presence felt at the hearings investigating Wheeler, usually in the traditional staff role of whispering quietly in a council member's ear. At one point, though, Evans engaged Wheeler directly in a lengthy exchange concerning Wheeler's alleged failure as a prosecutor. As Wheeler defended his record, Evans burst out, "In the prosecution of these aliens you did nothing! That's what I blame you for!"[33]

In Wheeler's final session before the Montana Council of Defense, his anger would pour out in succinct summary: "How would you like to have some little body of men get together in secret meetings and pass resolutions?" The council, said Wheeler, had done "everything possible in the world to prevent my reappointment in Washington, and it has been done for political reasons."[34]

It was the council, though, that would have the last word. Upon the conclusion of Wheeler's testimony, the council passed (again) its resolution: "The members of the State Council of Defense . . . desire to protest against the reappointment of Hon. B.K. Wheeler to the position now held by him of United States District Attorney for Montana."[35]

While the Montana Council of Defense busied itself at the state level, its county and local counterparts were equally active in communities throughout the state. When the legislature expanded the power of the state council, it also conferred new powers onto the subsidiary organizations. In essence, county and local councils of defense had the same power as their state counterpart—though the state council could overrule their decisions. In many ways, the county and local committees were more zealous than the state council, and "hundreds of Montanans found themselves jailed, humiliated or investigated."[36]

Local councils around the country were organized as "Liberty Committees," "Liberty Loan Committees," or "Third Degree Committees." One focus of local activity was the sale of Liberty Bonds, and in the early days of the war the instruments had provided an important voluntary mechanism for citizens to offer financial support for the war effort. As the war went on, though, the drive to sell bonds became more pointed. Quotas were established for the number of bonds that each person was expected to purchase, and criticism became sharp for citizens failing to buy their allotted share.

The term "slacker," which in the early days of the war was reserved for draft dodgers, began to be applied to anyone deemed not to have purchased enough bonds. Advertisements urging the purchase of bonds used language like "No mercy for bond shirkers." One ad in the *Seattle Post-Intelligencer* read, "A bond shirker is an enemy to humanity and liberty, a traitor and a disgrace to his country."[37]

With the power of investigation, council screeds against bond shirkers were not limited to generic advertisements. Many citizens were dragged in to testify. The formal interrogation of a Montana farmer named Victor Brown by the Stevensville War Service League gives a flavor of the harassment to which people were subjected:

QUESTION: *You own the property you are living on?*
ANSWER: *I have a deed; I will own it if I pay.*

QUESTION: *There is an encumbrance against it?*
ANSWER: *Yes, sir.*

QUESTION: *How many acres?*
ANSWER: *Well, it is a 320-acre tract.*

QUESTION: *How much did you pay for it?*
ANSWER: *Thirteen thousand dollars.*

QUESTION: *How much is against it now?*
ANSWER: *Nine thousand dollars.*

QUESTION: *Have you bought any Liberty Bonds?*
ANSWER: *No, sir.*

QUESTION: *Any war savings stamps?*
ANSWER: *No, sir.*

QUESTION: *Contributed anything toward the War Service League or Red Cross?*
ANSWER: *Nothing to speak of. My wife and I are members— we contributed our membership.*

QUESTION: *Any reason why you refuse to do so?*
ANSWER: *I have considerable debts to meet. While we are in entire sympathy with all the work in that organization we have felt we should meet our obligations . . .*

QUESTION: *Don't you think the country would be in a pretty bad way if only those who did not owe anything contributed towards the War Service League or bought bonds?*

Later in the testimony, the questioner asked if Mr. Brown *intended* to buy bonds.

ANSWER: *We are very glad to when we see ourselves clear.*

QUESTION: *In other words you don't feel you are able to do it until you pay all of your debts?*
ANSWER: *Not all of our debts. We deny ourselves a great many things we would like to have. We are living in a wreck of a three-roomed house. The improvements on that place are in bad condition and we don't feel able to pay six or seven dollars a day to a carpenter . . .*

ANSWER: *In other words you are looking forward to your own comfort at all times?*[38]

Farmer Victor Brown ultimately was allowed to leave, but only after the Stevensville War Service League had designated him as a "money slacker and as such deserving of public censure." Not all of those called before the Montana Council of Defense and its affiliates would be so lucky. Under the Montana Sedition Act, for example, forty-seven Montanans were imprisoned—some for sentences of twenty years.[39]

A s the months after the Extraordinary Session unfolded, Sam C. Ford found himself growing increasingly uneasy. Ford served as the attorney general of Montana and also as the official legal counsel to the Montana Council of Defense. A staunch Republican, he would one day be elected to the governorship.

By May of 1918, Ford had come to view the activities of the state and local councils of defense with increasing dismay. In his role of legal advisor, he had recommended against several actions contemplated by members of the Montana Council—telling them, for example, that they did not have the power to shut down the *Butte Bulletin*. Ford had been particularly dismayed by a Miles City incident in which a controversial speaker was denied the right to make a public address—then was beaten by a mob. Ford called it a "flagrant example of the very *kaiserism* against which the United States is warring today."[40]

On May 27, 1918, Ford felt that he could no longer stand idly by. He drafted a four-page "Communication from the Attorney General" and addressed it to the state legislature. "I feel it my duty to call to the attention of your honorable body recent violations of the state constitution and laws in a number of counties of the state." Ford described the Miles City incident, as well as numerous other cases in which the rights of free speech, free press, and free assembly had been repressed. He asked the legislature to rein in the defense councils "[i]n view of the participation of members of county councils in the lawlessness described."[41]

"It is true," wrote Ford, "that we are at war and that the life of the nation is at stake." But "it is also true that the primary purpose of this war is to uphold the fundamental principles of freedom and to prevent autocratic government, the rule of might, from being established on this continent."

Members of the Montana Council of Defense were stunned by Ford's letter, though they needn't have worried. His protest was a tiny ripple on the surface—a pebble in the sea. The great tide that floated the council, meanwhile, would continue to rise. Only two weeks before Ford submitted his letter, President Wilson had signed into effect the National Sedition Act.

The new national law was a virtual carbon copy of the Montana law. It was Montana's U.S. Senator Henry Myers, after all, who had written the original text on which the Montana law was based. Myers's bill had been stuck in the Senate Judiciary Committee since he introduced it in the immediate aftermath of Frank Little's murder. By the spring of 1918, though, federal interest in a stronger antisedition statute had been "whipped into a state of frenzy."[42]

Montana's two senators (Myers and Walsh) may have disagreed over the reappointment of B.K. Wheeler, but they found consensus on the need for a stronger federal sedition law. Together the two men would spearhead the successful effort to create a national version of the Montana law.[43] Both were motivated in large part by the lynching of Frank Little (along with Judge Bourquin's decision in the Ves Hall case). Without a federal sedition law, they argued, public frustration would again boil over in the form of mob violence. Though their logic would carry the day, it was an ironic rationale. Events in Montana, as state Attorney General Ford's letter made clear, showed that the sedition law did not *stifle* the rule of the mob—it *institutionalized* the rule of the mob.[44]

The National Sedition Act—with its direct lineage to events in Butte—stands as one of the most sweeping violations of constitutionally protected freedoms in modern American history, "a severe overreaction based on excessive and ill-informed fear." Two thousand Americans were prosecuted under the act for expressing political opinions deemed unlawful, and hundreds were sent to prison for terms of up to twenty years. An embarrassed U.S. Congress repealed the law on December 20, 1920.[45]

In a recent comprehensive analysis of wartime restrictions on free speech in America, constitutional scholar Geoffrey Stone went so far as to conclude that negative reaction to the World War I sedition law gave birth to the "modern civil liberties movement." In hindsight, "many supporters of the Wilson administration were shocked by what the nation had done."[46]

The fight over the future of B.K. Wheeler would continue after his investigation by the Montana Council of Defense. The council had recommended against the renewal of Wheeler's appointment as federal district attorney, but the decision remained with the White House. The White House, meanwhile, continued to wait for Montana's two Democratic senators to come to some type of agreement. Senator Myers had made his disapproval of Wheeler abundantly clear. For a while, though, Senator Walsh continued to offer guarded support for the man he had helped to place in office.[47]

The uncertainty continued through the summer and early fall of 1918. Will Campbell, of course, kept up his drumbeat in the *Helena Independent*. His editorials warned Senator Walsh that his continued support for Wheeler would undermine his own political future.

"Senator Walsh is going to be asked why men who preach seditious sermons in Iowa are promptly arrested by the United States district attorney while in Montana they have to be lynched."[48] The pressure on Senator Walsh would continue to mount. He did not want to stand against his friend and ally, but at the same time it appeared increasingly clear that support for Wheeler could cost him the November election.

In October, finally, a public clash between Wheeler and Anaconda would bring the matter to a head. A new miners' strike had broken out, and an Anaconda attorney named Dan Kelly charged that "federal officials" (a thinly veiled reference to Wheeler) were "counseling every day with the men back of the movement to curtail production." Wheeler fired off a letter to Kelly with a countercharge—that the IWW was encouraged to call the strike by "paid agents" of Anaconda. "If these facts are unknown to you," offered Wheeler, "may I suggest you consult the detective reports of your company."

Kelly responded with a letter of his own, complaining anew about Wheeler's handling of the Frank Little case. "Consistently you have sought to crucify patriotism and sanctify sedition in this state," he wrote. "Your picture hangs in the halls of this country's enemies." Both the Wheeler and Kelly letters were printed together in the *Anaconda Standard*.[49]

The day after the letters were published, Wheeler received a phone call from a Butte attorney named C. B. Nolan, a close political confidant of Senator Walsh. With the election by then only a month away, it appears clear that Walsh could no longer stomach a new round of bad publicity via Wheeler. "Do you plan to issue any more statements like that?" asked Nolan, a reference to the letter.

"As long as the Company keeps up its attack on me," replied Wheeler, "I will hit back with all the ammunition I have."

"Such statements won't help man, God, or devil," said Nolan. He then convinced Wheeler to travel with him to Washington along with the Montana State Democratic chairman. The intention of the trip, Wheeler knew, was to "kick me upstairs."[50]

In Washington, Senator Walsh came to Wheeler's hotel for a private meeting. "Well," said the senator, "I guess they will beat me for reelection if you continue as DA." On October 9, 1918, B.K. Wheeler issued a curt statement to the press. He was resigning, he said grudgingly, "in order to satisfy the friends of T. J. Walsh who believed my retention in office would mean his defeat as a candidate to succeed himself in the Senate."[51]

The Company-controlled press, including Will Campbell's *Helena Independent,* could scarcely contain its glee at Wheeler's demise. "The people of Montana know now after eighteen months' experience in a world war, what a real United States district attorney would mean to a state."[52]

Wheeler, meanwhile, returned to his private law practice in Butte. "My enemies really did assume my public career had ended."[53]

Twenty-five

•

"DOWN DEEP"

People are shockproof in Butte.
Down deep there's a certain optimism for the future.
—NEIL LYNCH, BUTTE HISTORIAN, MAY 1982[1]

The unruly braids of Butte history defy those who search for tidy summation.

For the players of the North Butte disaster, the final chapter was not written in 1917 or even in 1918. Burton K. Wheeler ultimately would carry the legacy of the great fire to the nation's capital, where his experiences would shape a pivotal battle with the president of the United States. As for the corporate actors, the fire would push the North Butte Mining Company to the brink of ruin, and though Anaconda in 1918 appeared to be as powerful as ever, its destiny would ultimately be written by forces even larger than the Company's colossal grasp. For the miners, representation by a meaningful union would remain elusive until the landmark changes of the New Deal. And as for Butte itself, not even the New Deal could bring security to a town built atop the shaky foundation of copper.

O thers in B.K. Wheeler's situation might have faded quietly into historical obscurity. He had fought hard against steep odds, but in 1918 it appeared that Anaconda and Will Campbell had prevailed. Wheeler, at least, had kept his independence and his principles intact. With a young family, a comfortable income, and a vacation house on Lake McDonald, there must have been a temptation to let others carry the fight.

Wheeler, though, put his bitterness to work. In 1920, he ran for governor of Montana on an explicitly anti-Company platform. His campaign pledge: "If elected I will not put the Anaconda Copper Mining Company out of business, but I will put it out of politics." Anaconda fought back with all of its considerable power, dubbing Wheeler "Bolshevik Burt" for his anti-Company views. (The riposte of Wheeler's supporters: "We are opposed to private ownership of public officials.")[2]

The irony of Montana's 1920 gubernatorial election was that *both* candidates were anti-Company. Wheeler's Republican opponent— Joseph Dixon—had run Teddy Roosevelt's progressive "Bull Moose" campaign in the state. Anaconda, forced to choose between the lesser of two evils, put its political muscle behind Dixon. Dixon's tough line on the Company, meanwhile, peeled away votes from Wheeler's core constituency. Dixon won the election in a landslide.[3]

Though Wheeler could not know it as the 1920 returns were counted, his loss was a blessing in disguise. For Montana, the Great Depression began in the immediate aftermath of World War I—a full decade before the rest of the country. Farm prices plummeted while at the same time drought plagued the state. Half of Montana's farmers lost their land. Half of all banks failed, swallowing family savings. Tens of thousands of people fled the state for the brighter prospects of the West Coast.[4] (This was the era so evocatively painted in Wallace Stegner's *Big Rock Candy Mountain*.) As for Butte, the end of the war cut demand for copper, resulting in the shutdown of several mines. For those miners still working, Anaconda cut wages by $1 per day.[5]

It was a terrible time to govern, and Governor Dixon would pay the price. Voters threw him out of office after one term. B.K. Wheeler, meanwhile, was unrelenting. In 1922, Montanans sent him to the United States Senate—launching a twenty-four-year career that would establish him as one of the most powerful men of the era.[6] As in Montana, Wheeler's trademark in Washington would be his stubborn independence. Were it not for this characteristic, events would show, the Butte attorney might well have been president of the United States.[7]

In Washington, Wheeler made an immediate mark—finding a Harding administration rife with corruption. In 1924, Wheeler led the charge against Harry M. Daugherty, the attorney general of the United States. Weeks of high-profile hearings turned up a complex web of misdeeds, and Daugherty was ultimately forced to resign.[8] Wheeler emerged from the hearings with a national reputation as both a crusader and a maverick, a status he buttressed further when, in the fall of 1924, he ran as Robert La Follette's vice presidential candidate on the Independent Progressive ticket.[9]

One of the more controversial issues in the 1924 election concerned the Ku Klux Klan—then at the height of its national power. The Democratic National Convention rejected a platform plank condemning the Klan, and Republican nominee Calvin Coolidge remained silent on the issue. When La Follette and Wheeler denounced the KKK, the Klan made the defeat of the two men an overarching national goal. Wheeler would require special protection while campaigning in the South. The La Follette-Wheeler ticket ultimately earned 17 percent of the 1924 vote. (For comparison, Ross Perot in 1992 won 19 percent.)[10]

Though the 1920s would witness Burton K. Wheeler's rise as a national figure, far greater drama lay ahead. It was the decade of the 1930s that would provide the backdrop to what Wheeler described as his "long and bumpy political relationship with FDR."[11]

B.K. Wheeler was the first U.S. senator to advocate Franklin D. Roosevelt for president. Even before the 1932 Democratic National Convention, Wheeler was busy lining up prominent supporters, including Boston financier Joseph P. Kennedy and Louisiana kingpin Huey Long. After FDR's victory, there was talk that the president might name Wheeler to the post of attorney general, but Wheeler claimed to prefer the independence afforded him by Congress. "[B]eing in line for the chairmanship of the Interstate Commerce Committee," he explained, "I had much more to gain by remaining in the Senate."[12]

Wheeler emerged as one of Roosevelt's key Senate allies in pushing the great raft of New Deal legislation through Congress. So close was Wheeler's early relationship to the president that Senate colleagues called him "teacher's pet." FDR rewarded Wheeler's loyalty, in classic fashion, with the construction of the gigantic Fort Peck Dam on Montana's Missouri River.[13]

This close relationship between Wheeler and Roosevelt would extend (with occasional blips) through the president's first term. Indeed the great friction between Wheeler and Roosevelt would not arise in the context of executive relations with the *legislative* branch, but rather in the president's relationship with the *judiciary*. While Congress was a ready partner in FDR's New Deal, the Supreme Court was not. Four conservative members—known by the apocalyptic appellation of the "Four Horsemen"—began to strike down various New Deal components as unconstitutional infringements on personal liberty. FDR's great legislative achievements began to crumble.[14]

The president, after the mandate of a landslide victory in 1936, prepared to strike back. His actions of 1937 are perhaps the most controversial of his presidency. In the center of the controversy would stand B.K. Wheeler, and central to Wheeler's motivation were lessons he learned in the tumultuous aftermath of the North Butte disaster.

The New Deal helped put Montanans back on their feet. Indeed, on a per capita basis, Montana ranked second among all states for New Deal investments. Agricultural programs (especially the Agricultural Adjustment Act, or "AAA") raised and stabilized farm prices. The Silver Purchase Act pushed up the price of silver. The Civilian Conservation Corps put thousands of young men to work on jobs ranging from fighting forest fires to building campgrounds. The Fort Peck Dam project—FDR's patronage to B.K. Wheeler—employed 10,500 workers at the peak of its construction in 1936.[15]

For Butte, though, no New Deal program had more significance than a set of new laws protecting the rights of workers and unions. In 1933—for the first time since the heady, early days of the 1917 strike—it appeared that the miners stood a fighting chance against the power of Anaconda.

The period between 1917 and 1933 had seen a continuation of the grim losing streak for miners and unions in Butte. Anaconda, of course, had proved itself fully capable of taking down strikes and other forms of dissent by miners. Then the World War I sedition statutes would add still more weapons to the Company arsenal. In August 1918, the entire IWW leadership was tried and convicted under the federal Espionage Act at a trial in Chicago.[16]

In Butte, the end of the war resulted in "practically no demand for copper," and Anaconda closed several of its mines and reduced wages. Adding to an oversupply of labor was the return of soldiers from the war in Europe seeking to resume their work in the mines. Strikes broke out in 1918 and 1919, but neither produced significant results for the miners. Despite the crackdown on the IWW, the radical union now played the explicit leadership role in the strikes—in large part because no other meaningful organization existed.[17]

Matters finally came to a head in April 1920. The IWW called yet another strike, establishing pickets on the main roads to the mines. On the afternoon of April 21, the local sheriff and his men attempted to defuse a tense standoff between armed Company guards and a large group of unarmed miners. In the midst of the sheriff's effort to sort out the situation, the Anaconda gunmen opened fire. Sixteen miners were shot—all in the back—as they attempted to flee. One of the men would die the next day.[18]

The sheriff and his men testified that all the shots had come from the Company gunmen, though they could not be certain who *specifically* had shot whom. No one was ever charged, and the miners called off the strike a few weeks later.[19] The 1920 walkout would be the last significant labor upheaval for more than a decade.

On June 16, 1933—the final day of Roosevelt's famous "Hundred Days" legislative session—Congress passed a law known as the National Industrial Recovery Act (NIRA). To promote industrial recovery, NIRA allowed companies in the same industry to engage in certain activities that would otherwise have been illegal under antitrust laws (for example, establishing production quotas among companies and fixing prices). There was a catch, though. Companies benefiting from the new law would also be required to give workers "the right to organize and bargain collectively through representatives of their own choosing." The law also forbade employer interference or coercion.[20]

For the nation's downtrodden unions, the effect of the new law was profound—a signal that the weight of the president of the United

States now stood behind them. All over the country, "[u]nions used it as a basis for telling unorganized workers in organizing campaigns that President Roosevelt wanted them to join unions." By the end of the summer of 1933, scores of strikes had broken out nationwide, forcing the president to establish a National Labor Board to enforce the right to organize and to mediate disputes.[21]

In Butte, miners began an immediate effort to revivify the old Butte Miners' Union, once the foundation on which "the Gibraltar" was formed. In the summer of 1933, an organizing drive signed up 4,500 new members. With a mandate based on swelling membership and the implicit encouragement of the president of the United States, the Butte Miners' Union demanded negotiations with Anaconda. Talks began in April 1934, but the Company steadfastly refused the miners' demands for higher wages, a forty-hour workweek, and most importantly, recognition of the union itself.[22]

On May 8, 1934, Butte miners walked off the job. At first, it appeared that Anaconda would simply pull out the old playbook, running the same tried-and-true strategy that had successfully crushed strikes for two decades. The Company's domination of Montana's press, for example, was still alive and well in the 1930s, and the papers excoriated the strikers. "There is little sympathy in the community with the present move to close down the industry on which the town depends," editorialized the *Butte Daily Post*.[23]

The refrain sounded eerily familiar. The times, though, had changed, and it was the Company that was now out of tune. Federal mediators traveled to Butte and helped organize a series of negotiations. The miners held firm. And this time, unlike 1917, the goal of federal mediation was not the immediate resumption of copper production, but rather a sustainable reconstruction of the relationship between workers and management.[24]

Back in May, a member of the machinists' union named James

O'Brien had predicted a long strike, with no resolution "until snow falls." On September 19, 1934, beaming union representatives marched into Anaconda headquarters at the Hennessey Building. The agreement they signed gave them a wage increase, the return of a forty-hour workweek, and formal recognition of the union—including a closed shop. As the men put their pens to paper, snowflakes began to fill the air.[25] Snow had come early, but for the miners of Butte, a winter stretching two decades had passed.

B urton K. Wheeler was in New York on February 5, 1937, when he read a newspaper account of a new plan announced by President Roosevelt. The president was asking Congress to enact a sweeping "reorganization of the judiciary." Under FDR's proposal, the president would be given the power to appoint one new Supreme Court justice for every sitting justice who remained on the bench beyond his seventieth birthday. Six of the nine justices were then over seventy. If Congress accepted the proposal, Roosevelt would have the immediate power to "pack the court" with his own judges. His goal, of course, was to stack the deck in his favor.[26]

Wheeler's reaction was immediate. "Here was an unsubtle and anti-Constitutional grab for power which would destroy the Court as an institution," he said. "I felt I would have to do everything I could to fight the plan."[27] Over the next five months, congressional opponents of the court-packing plan would battle one of the most powerful presidents in history. B.K. Wheeler would lead them.[28]

FDR would apply every ounce of both his silk-glove charm and his brass-knuckle politics. At one point, the president sent labor leader Sidney Hillman to lobby Wheeler in support of the plan. In a revealing glimpse, Wheeler explained to Hillman that his unyielding opposition to court packing was founded on his own experience as a target of abusive power in Montana during World War I. "Another hysteria might sweep this country," said Wheeler, "and it might be against you people, or some other group, and when that time comes they will be looking to the Supreme Court to preserve their rights and uphold the Constitution."[29]

The echoes of Montana in 1917 and 1918 are also apparent in a Wheeler radio address attacking the president's plan. "Create now a political Court to echo the ideas of the executive and you have created a weapon; a weapon which in the hands of another president could . . . extinguish your right of liberty of speech, or thought, or action, or of religion."[30]

So radical was the court-packing plan that Wheeler found allies among the Supreme Court's most liberal justices—the same men who had been fighting against the "Four Horsemen" in a losing effort to uphold New Deal legislation. Particularly important was Wheeler's close personal relationship with Justice Louis Brandeis, whom a Brandeis biographer described as Wheeler's "father confessor since Wheeler's first years in Washington."[31]

At the apex of the debate over the court-packing issue, Justice Brandeis's wife paid a visit to Wheeler's daughter Elizabeth, who had just given birth. As Mrs. Brandeis was leaving, she remarked to Elizabeth, "You tell your obstinate father we think he is making a courageous fight." When Elizabeth conveyed the message to Wheeler, he interpreted it—correctly—as a deliberate signal that Justice Brandeis opposed FDR's court reform proposal. Wheeler immediately called on the justice.[32]

In a private meeting at Brandeis's home, Wheeler asked him if he and/or Chief Justice Charles Evans Hughes would testify before Congress in opposition to the court-packing plan. Brandeis replied that he did not believe such testimony was appropriate, but, he continued, "You call up the chief justice and he'll give you a letter."[33]

"I don't know him," protested Wheeler.

Brandeis immediately picked up the phone, called the chief justice at home, and told him that Wheeler needed to see him. Hughes asked Wheeler to come over at once. The next day, a Sunday, Hughes enthusiastically provided Wheeler with a letter.

On Monday morning, Wheeler was scheduled to testify on the court-packing issue at an important hearing of the Senate Judiciary Committee. After discussing his own opposition to Roosevelt's plan, Wheeler reached dramatically into the breast pocket of his suit. "I have a letter by the Chief Justice of the Supreme Court, Mr. Charles Evans Hughes, dated March 21, 1937, written by him and approved by Mr. Justice Brandeis and Mr. Justice Van Devanter." Wheeler then read the letter to a spellbound audience, systematically undermining the president's key arguments. As Wheeler remembered it in his autobiography, "You could have heard a comma drop."[34]

The effect of the chief justice's letter was electric, and at least one member of the Roosevelt administration believed that it "did more than any other single thing to turn the tide."[35] On May 18, 1937, the Judiciary Committee voted 10–8 against the FDR bill to pack the court. Leaving no room for ambiguity, the committee's report concluded with the following appraisal: "It is a measure which should be so emphatically rejected that its parallel will never again be presented to the free representatives of the free people of America." In July, the full Senate voted the bill down, 70–20.[36]

For Wheeler, it was a moment of personal and political triumph. Within a few short years, though, Wheeler would again oppose FDR—this time on the momentous issue of war. In this new fight, Wheeler's judgment would fail him, with consequences that would end his career.

B y 1939, all of Europe was at war, and by 1940, Nazi Germany occupied most of the European continent. In the years before the outbreak of war, the world had watched as Adolf Hitler took increasingly aggressive actions—including the beginnings of the Holocaust.

Like many Americans of the era, Wheeler drew false analogies between World War I and World War II. The First World War had left searing scars on the nation, and certainly on B.K. Wheeler. "I wanted to see the American people keep their heads this time," said Wheeler.[37]

As the events of the 1930s unfolded, though, Wheeler would instead establish himself as a paragon of head-in-the-sand isolationist belief. Unlike World War I, the origins of this new global conflict did not lie in the obscure, secret diplomacy of Europeans—but rather in the explicit expansionist and totalitarian aims of Germany and Japan. Though Wheeler publicly condemned Hitler, he mistakenly believed that Germany did not threaten the United States. "This was not our war," said Wheeler in the years before Pearl Harbor.[38]

A major flashpoint centered over FDR's 1940 proposal for the Lend-Lease program. Under Lend-Lease, the United States would provide military equipment to Great Britain and Russia. America, urged the president in a fireside chat, must be the "arsenal of democracy." Wheeler, in a rebuttal broadcast on national radio, called the Lend-Lease proposal "the New Deal's triple-A foreign policy" (a reference to the farm program, AAA), stating, "it will plow under every fourth American boy."

Roosevelt was outraged. Wheeler's comment, he said at his next press conference, "was the most dastardly, unpatriotic thing that has been said in public life in my generation. Quote me on that."[39]

Only six months earlier—in the run-up to the 1940 election—Roosevelt had nearly selected B.K. Wheeler as his vice presidential running mate.[40] Wheeler was a logical candidate, despite his opposition to FDR on the Supreme Court and the war. In fact, Wheeler believed that Roosevelt was considering him precisely *because* of his opposition to the president on those issues. "FDR ever since the Court fight considered me an opponent to be reckoned with and doubtless felt that relegating me to vice presidential 'limbo' would neutralize me, so to speak." B.K. Wheeler had no interest in any position that would neutralize him. At a Washington dinner party, the senator was asked why he wouldn't consider the nomination. "Because the president is going to get us into the war," said Wheeler, "and I won't go out and campaign and say he won't."[41]

Had Wheeler become Roosevelt's vice presidential nominee, there is a significant possibility that the Montanan would have become president of the United States. FDR's 1944 running mate, Harry S Truman, would assume the presidency when Roosevelt died in office on April 12, 1945.[42]

As for Wheeler, he did come out strongly in support of the war after Japan's December 7, 1941, attack on Pearl Harbor. (Montana's representative in the House, the diehard pacifist Jeanette Rankin, cast the lone vote against the declaration of war.) Wheeler's past advocacy of isolationism, however, would be proven misguided by the events of World War II. In the 1946 election he failed even to win the primary. After twenty-four years in the Senate, Wheeler retired from public life and spent the rest of his career in the private practice of law.[43]

Burton K. Wheeler's life inspired at least two works of fiction. His investigation of Harry Daugherty—the corrupt attorney general—was the inspiration for a film script called *The Man from Montana,* which was produced in 1939 under a different title—*Mr. Smith Goes to Washington.*[44] More recently, Wheeler served as the sinister foil in Philip Roth's fictional "alternate history," *The Plot Against America.*

Neither portrait provides much insight into Wheeler, a man whose defining contours had been carved by his formative experiences in

Butte. Far from the naive Mr. Smith, Wheeler arrived in Washington as the thick-skinned survivor of bare-knuckle politics in an utterly cut-throat era. By the same token, though, surely Roth could have found a more likely prototype than Wheeler for his plotting American tyrant. In Wheeler's long public career, he had fought for the rights of power-less workers against an all-powerful corporation; condemned a World War I witch hunt in the face of enormous public pressure; rooted out official corruption in the highest levels of the federal government; stood up to the KKK when neither major political party would do so; and—most famously—preserved the independence of the Supreme Court as a bulwark against the oppression of minorities, an action motivated by his own experience as a victim of dictatorial power.

All in all, in the words of Wheeler, a life "as full as I could wish."[45]

Readers of the *Wall Street Journal* on July 27, 1923, might easily have missed the two-paragraph story tucked into page three under the simple headline "North Butte." The short piece described the end of operations at the North Butte Mining Company. "All pumps, rails and other equipment were hoisted to surface Tuesday from the Speculator and Granite Mountain mines." With the pumps gone, the mines would flood. "Three watchmen are now the only employees."[46] Between the fire of 1917 and the economic disaster of the early 1920s, the North Butte had been unable to survive.[47]

The Anaconda Copper Mining Company, by contrast, seemed "too big to fail." It would bend but not break beneath the same brutal forces that took down the North Butte. After surviving Montana's depression of the 1920s, it would go on to survive the Great Depression of the 1930s. It was not until the 1940s, though, that Anaconda would soar again, as World War II transformed American industry into the "arsenal of democracy."[48]

Even as an economic depression and two world wars shaped the landscape, the copper industry felt the leading edge of a phenomenon that would take another half century to unfurl: the global economy. Indeed Butte, Montana—two oceans and half a continent removed from the rest of the world—would be buffeted by global forces long before those forces became "globalization."

Anaconda saw the trend coming early, responding in two ways that would shape the future of Butte. In 1923, Anaconda bought Chuquicamata—the Guggenheims' famous Chilean copper mine—for $77 million, then the largest cash deal in Wall Street history. Chile was the "Saudi Arabia of copper," and Anaconda intended to own its share. For fifty years, the Company poured hundreds of millions of dollars into ever-expanding facilities, and by the 1970s, "Chuqui" would generate as much as 75 percent of Anaconda's annual profits.[49]

Though Chile was now the center of Anaconda's business universe, enormous copper reserves still remained in the (former) "richest hill on earth." The problem in Butte, though, glaring by the late 1940s, was that the remaining copper was low grade. Mining it using the old methods of tunneling below ground could no longer pay for itself.[50] The solution—the open pit—had been available for decades. But the assumption had always been that open-pit techniques could not be applied in Butte because of, well, Butte. It was one thing to have headframes, slag heaps, and smelters cropping up among the buildings. It was quite another thing to dig up the whole town.

In 1955, though, facing the loss of its lifeblood, Butte made the only deal that it could. To keep the mining on which the city depended, the city would become the mine. It was the birth of the Berkeley Pit—the physical feature that today defines the town.

It is hard to describe the pit to someone who has not seen it. Nor do photos begin to hint at its awesome scale. In the 1950s, Anaconda paid a prominent business historian, Isaac F. Marcosson, to write the

Company history. In his book, *Anaconda,* Marcosson described the open pit mine at Chuquicamata. "To view its massive proportions," he wrote, "is to recall instinctively Kipling's words: 'They shall splash on a ten-league canvas with brushes of comet's hair.' "[51] It is doubtful that anyone but Marcosson was ever thus inspired at the sight of an open-pit copper mine.

The term "pit" fails to capture the awesome destruction, implying the type of hole you might dig with a shovel. The Berkeley Pit, by contrast, is a mile and a half long, a mile wide, and more than a quarter mile deep, with terraced walls designed to support the weight of 200-ton trucks.

The scale of the Berkeley Pit is perhaps best understood not in a description of what it *is,* but of what it *replaced.* In almost three decades of operation, the Pit swallowed entire neighborhoods—fifty square blocks, including Dublin Gulch, Meaderville, and Fintown. Lost too were churches, cemeteries, and the sixty-eight acres of Columbia Gardens. The homes of *six thousand people* were destroyed. The Company was required to—and by all accounts did—give fair compensation for what was taken, at least so far as what was taken could be valuated in dollars. There was no alternative to leaving, in any event. Under a nineteenth-century law, Anaconda had the right—normally reserved to sovereign powers—of eminent domain.[52]

The Berkeley Pit bought Butte twenty-seven more years of mining.

In 1970, the people of Chile elected as their president a socialist physician named Salvador Allende. The following year, Allende nationalized the Chilean copper industry, including Anaconda's Chuquicamata. Overnight, the action wiped out Anaconda's most valuable asset—and the source of three-quarters of its profits. The Company staggered on the brink of bankruptcy, surviving only by selling off huge chunks of assets—its zinc operation, its timberlands. To cut expenses, hundreds of jobs were eliminated in Butte and elsewhere.[53]

Hope revived in 1977, when the Atlantic Richfield Company—a cash-flush energy conglomerate known as ARCO—bought the remnants of Anaconda. Global trends, though, would continue to conspire against Butte. Unionized American copper had difficulty competing against low-cost suppliers in South America and Africa. Meanwhile, copper itself was under pressure, increasingly replaced by fiber optics and satellite transmission. Copper prices plummeted.[54]

The ax fell close to Butte when ARCO, in 1980, shut down the giant smelter in the town of Anaconda, twenty-six miles away. In Butte, where rumors about the smelter had been rampant, local leaders pressed for assurances that ARCO would at least continue to operate the mines. Ralph Cox, ARCO vice president, assured Don Peoples, Butte's tough-minded chief executive (mayor), that regardless of the verdict on the smelter, copper mining would continue in Butte "for several decades." The Berkeley Pit, said Cox, had at least twenty to thirty years of life.[55]

On a snowy day in January 1982, Chief Executive Peoples was driving back to Butte from Seattle when he stopped in Coeur d'Alene to call his secretary. She was frantic. "ARCO people have been trying to get ahold of you all day," she said. "They have an announcement—the rumor on the street is that they're going to close the mining operation." By the time Peoples could return to Butte, the announcement had been made. In hindsight, he admits, "the writing had been on the wall for a long time." Still, it is difficult to overstate the shock. "It was almost like a death in the family."[56]

A mericans at the turn of the twenty-first century are accustomed to viewing the hollow vestiges of cities and towns left back in the wake of time, but there is nothing like Butte. The remnants of the town—including the old Granite Mountain headframe—perch precariously on the edge of the Berkeley Pit. The Pit itself is now part of the largest Superfund site in the country. When ARCO stopped mining, it also stopped pumping the constantly rising groundwater from the depths of the pit. As a result, water has flooded in, mixing with a century's worth of mine and smelter tailings. A highly polluted creek also dumps toxic water into the Pit.

Postcards of the Berkeley Pit show its water as indigo blue, like some pristine mountain lake. In reality it is a rust-colored stew of deadly chemicals, including arsenic, copper, and cadmium—potent enough to melt metal. In one infamous episode, a flock of snow geese landed in the Pit, dying there when they sipped its water. *And the water is rising.* Without intervention, the poisonous water would spill over the edge of the Pit around 2025—flooding the town and the valley below.

There is no solution that will make the problem go away. "The magnitude of the problem is so enormous," says Tracy Stone-Manning, a local environmental leader. "The best we can do is figure out an OK Band-Aid." The current idea is to pump and treat the water for the rest of time, holding the perpetually rising tide below a critical level.[57] One piece of the plan, a water-treatment facility, recently came online. Butte residents monitor the status of the pit on a popular local Web site, www.pitwatch.org.

In May 1982, four months after ARCO announced the shutdown of mining in Butte, the *Montana Standard* (formerly the *Anaconda Standard*) ran an article with the brave-faced title "It's Still the 'Richest Hill on Earth.'" In the article, a local historian was asked his view of the shutdown. "People are shockproof in Butte," he said. "Down deep there's a certain optimism for the future."[58]

In the years that followed there have been reasons for such feeling. From a symbolic standpoint, there may be nothing more significant than the resumption of mining. In 1986, a Montana entrepreneur named Dennis Washington bought the remnants of Anaconda from ARCO for pennies on the dollar. Today he employs nearly 400 miners at a highly efficient open-pit operation just east of Berkeley. In what is still a union town, there seems to be surprisingly little residual resentment that the new jobs are all nonunion. People are just glad to see mining return to the Mining City.[59]

Beyond mining, there are other signs that point to a longer-term job base for Butte. A local development authority has used considerable creativity to attract high-tech manufacturing, including a company called Advanced Silicon Materials—"ASMi." ASMi built and operates a poly-silicon processing plant that employs 200 people, many of them engineers who graduated from Butte's Montana Tech University. Don Peoples, the mayor of Butte at the time ARCO shut down the mines, now heads a company called MSE Technology Applications, whose 200 employees conduct advanced research for clients including NASA, and earn wages that average $53,000 a year.

Despite these important success stories, it would be naive to claim glibly that Butte has turned the corner. In 2003, Butte suffered a blow that some feel was even more devastating than the 1982 shutdown of the mines. Since 1912, Butte had been headquarters to the Montana Power Company. Though legally independent, Montana Power and Anaconda for decades shared executives as well as legal and public relations staff. Many referred to Montana Power as Anaconda's "Siamese twin"—an integral part of the Company.[60]

After the mines closed, many in Butte consoled themselves with the thought that "we still have Montana Power." Unlike the boom-and-bust copper industry, Montana Power was a highly regulated utility company—historically among the most stable forms of enterprise. Generations of Butte citizens had bought Montana Power stock for its reliable, steady returns.

In the heady markets of the go-go 1990s, though, Montana Power executives decided that strict regulation was stifling the company's growth. They lobbied the state legislature, successfully, for deregulation. Freed of their former strictures, management decided to focus Montana Power in a new direction—the growth industry of telecommunications. The company changed its name to Touch America, and over a two-year period sold off a stunning $2.1 billion worth of "old economy" assets—dams, power plants, transmission lines, oil, gas, and coalfields. With the proceeds, Touch America bought fiber-optic telecom lines.[61]

For a while the strategy seemed brilliant. By the spring of 2000, the stock had risen from a steady price of around $30 per share to a high of $65. Then came the seemingly inevitable crash. When it was over, two years later, the stock sold at three cents. Bankrupt, the company was liquidated, though not before four senior executives took out $5.4 million in golden parachutes. Hundreds of Montanans, mostly in Butte, lost their jobs. Thousands of investors, including Montana Power pensioners, lost their life savings.[62]

Thus ends the latest chapter in the saga of Butte, Montana. Today Butte is a living archive of heroic and cautionary tales, but there is no happy ending to date. Which is why, to an outsider, one of the more striking attributes of Butte's people is a fundamental lack of self-pity— especially noteworthy in today's victimhood society. Certainly an incident like the Montana Power debacle sparks a firestorm of emotion, but people in Butte seem anxious to move beyond the bitterness and anger.

A Butte man named Jim Keane worked in the underground mines

as a summer job during college. Today he is the business manager for the local branch of the International Union of Operating Engineers and a Butte representative in the state legislature. "It's a huge set-back," he says when asked about the fallout from Montana Power. "But what do we get by sitting around and crying about it?"[63]

•

EPILOGUE
"NORMAL FOR ITS TIME"

The story of Butte in 1917 was altogether normal for its time.
Indeed, in that very normality lies the story's significance.
What took place in Butte took place elsewhere as well.
When we know the Butte story we know the others.

—REPORT PREPARED FOR PRESIDENT FRANKLIN D. ROOSEVELT, 1934

How do you sum up the North Butte disaster, let alone a century and a half of Butte, Montana? What meaning can be found today in the heroism of Manus Duggan, the resilience of Madge Duggan, or the relentless drive of B.K. Wheeler?

In the 1930s, a Justice Department official who studied the fire and its aftermath declared that "[t]he story of Butte in 1917 was altogether normal for its time. Indeed, in that very normality lies the story's significance. What took place in Butte took place elsewhere as well. When we know the Butte story we know the others."[1]

Certainly the story of the disaster speaks to an experience far broader than a single place or even a single time. Still, "normal" seems an unlikely descriptor. Far from normal, everything about Butte looms larger than life, magnified, exaggerated, like some gross caricature of itself. Butte's villains are more villainous, its heroes more heroic, its wealth more extravagant, its poverty more grinding. Butte's triumphs are the stuff of legends. Butte's tragedies are almost too painful to bear. And yet, in this caricature we see somehow more clearly—the essence standing out in stark relief.

As with most of American history, we are struck by the very nearness of these events. The history in Butte is not ancient or even particularly old, barely dusted by time. We can touch it, if we reach, across the span of a single life.

One of the investors who held on to Montana Power/Touch America stock, watching as it tumbled to worthlessness, is an eighty-eight-year-old retiree in Fort Worth, Texas. Her name is Manus Dugan Banko, the daughter of Manus and Madge Duggan. "I kept holding on as it went down," she said. "It was such a good company."[2]

Manus Dugan Banko was born a month after the death of her father, but she keeps him alive through the stories she heard growing up. She remembers, in particular, a story Madge used to tell about a day five months after the North Butte disaster. Madge stepped onto a Butte streetcar that afternoon carrying her four-month-old baby girl, swaddled all in white. A shift had just ended at the mines, and the car was crowded with men coming off work. One of the miners recognized Madge and knew that the redheaded baby was the child of Manus Duggan. The men nudged one another and stole shy glances. Finally one of them asked to hold the baby, reaching out with his work-stained hands. Madge handed the infant to the miner. Soon others wanted a turn, and Madge watched as her child—Manus's child— was passed from miner to miner. When the last man handed the baby back, her white clothes were black from their hands.

The story of the American West is the story of hope. Hope, defiant, in the face of overwhelming proof that it should not exist. Hope, even in Butte, where after every crushing blow, the people stand up again, dust themselves off, and go back to work. Hope so irrational that it can only be understood as faith.

ACKNOWLEDGMENTS

Dozens of Butte citizens (past and present) contributed their time and expertise to my research, including Harp Cody, Eddy Drabendt, Jim Edwards, Dick Gibson, Jim Harrington, Al Hooper (who at age ninety-four may be the last living person with personal memories of the disaster—he was six in 1917), Jim Keane, Don Peoples, Sally Wolahan Rasmussen, Mark Reavis, Pat and Kitte Robins, Kevin Shannon, Robin Urban, Jeanne Weber, and Carol Williams.

A particular thanks to Manus Dugan Banko, the daughter of Manus and Madge Duggan, who was kind enough to speak to me about this project on three different occasions.

Anyone intrigued by the events in this book should visit Butte, which may have more history per square inch than any place in the country. When you go, eat a pasty at Gamers Café. You can even spend the night at Copper King William Clark's former mansion— now a bed-and-breakfast.

Books like mine could not be written without the careful preservation of history in archives and libraries. The tireless Ellen Crain at the Butte-Silver Bow Archives is, quite literally, the font of all Butte knowledge. Ellen has two knowledgeable and helpful assistants: Shain Wolstein and Judy Strand. Thanks to Dick Gibson, Robin Urban, Susan Mattson, and Marie Kraus—who gave me access to the collections (including the photo archives) at Butte's fantastic World Museum of Mining. Thanks also to Brian Shovers and Rich Aarstad at the Montana Historical Society Library in Helena, and to George Burkes with the Congressional Research Service.

I am blessed with friends and family who are unwavering supporters as well as thoughtful readers. Their general encouragement and/or careful editing of my manuscript helped to make this book stronger. Thanks especially to Max Baucus, Mike Bridge, Sean Darragh, Ken Doroshow, Liz and John Feldman, Phil Gardner, Cheryl and Brent Garrett, Pam Kurland, Liz and Eric Jacobsen, Mickey Kantor, Jen Kaplan, Carol and Ted Kinney, Peter and Shelley Lambros, Amy and Mike McManamen, David Marchick, Randy and Julie Miller, Lori and Tim Punke, Marilyn and Butch Punke (who also helped me do research in Butte), Brent and Jo Ruby, Brian and Jo May Salonen, Peter Scher, Peter Stark, Mary and Bill Strong, Pennie and Gary Tague, Amy and Kevin Teague, Kim Tilley, Lynne and Gary Willstein, and Jay Ziegler.

Thanks to Allen Fetscher, Bruce Bugbee, and Bev Whitt for giving me a place to write as well as jolly companionship on a day-to-day basis.

I owe an enormous and growing debt to my agent, Tina Bennett at Janklow & Nesbit, whose support for my writing has given me the opportunity to live again in the West. Tina is assisted by the ever-helpful Svetlana Katz. At Hyperion, I'm grateful to Bill Strachan, my insightful editor, and also to Gretchen Young, Ruth Curry, and Allison McGeehon. Thanks also to Keith Redmon at Anonymous Content.

One of the great joys in writing this book was the exploration of

Butte with my children, Sophie (eight) and Bo (five). My initial interest in the North Butte disaster was triggered, in no small part, by a visit we paid to the World Museum of Mining. Sophie later helped dig through archives in both Butte and Helena. And both kids joined me for the panoramic views awaiting those who brave the bus ride up to Our Lady of the Rockies.

Thanks, finally, to my wife, Traci, who in an act of no small faith has come along for the ride.

NOTES

The author invented no dialogue in this book. The most common source for dialogue is quotations from contemporaneous newspaper interviews with participants in the actual events. (Butte in 1917 was served by three daily papers.) Other sources include author interviews, the coroner's inquest, and written accounts by participants. Specific sources are cited below.

Chapter 1. "There Is a Sign"

1. Descriptions of the safety work on June 8, 1917, are drawn from a Department of Interior/Bureau of Mines Report on the North Butte Disaster: *Lessons from the Granite Mountain Shaft Fire, Butte (Bureau of Mines Accident Report)*, 1922, pages 14–15, 31. The report was authored by Daniel Harrington, a Bureau of Mines employee who was on the ground (and underground) in Butte before, during, and after the disaster.
2. Ibid.
3. Antiwar handbill, June 4, 1917, on file at the Montana State Archives in Helena.
4. See Donald A. Garrity, *The Frank Little Episode and the Butte Labor Troubles of 1917*, unpublished thesis, April 1957, University of Montana Collection.
5. Antiwar handbill.
6. *Butte Miner*, June 11, 1917, page 8.
7. Burton K. Wheeler, *Yankee from the West* (1962), page 136.
8. *Anaconda Standard*, June 6, 1917, pages 1, 10; *Helena Independent*, June 6, 1917, page 1. Though the precise status of the troops is not clear, it appears they were recently federalized members of the Montana National Guard, on "utility duty" (guarding the mines) since the U.S. entry into the war.
9. Wheeler, page 137.

10. The estimate of the cable's cost appears in the *Butte Miner,* June 10, 1917, page 10. According to U.S. Bureau of Labor Statistics cited by Arnon Gutfeld in *Montana's Agony* (1979), page 18, the average annual wage of a Butte miner in 1917 was $1,215.

11. The official estimate of the number killed comes from the *Bureau of Mines Accident Report,* page 21. The actual number of men killed may be higher. Documentation at the time was sporadic, and the official coroner's inquest is missing. A Butte historian, James D. Harrington (no relation to Bureau of Mines Official Daniel Harrington), did an analysis of all available documentation and concluded that "at least 167" men died. See "A Reexamination of the Granite Mountain-Speculator Fire," *Montana,* Autumn 1998, pages 62–69. Erring on the side of inclusion, a memorial in Butte to the dead miners lists 168 names.

Chapter 2. *"Like a Gigantic Torch"*

1. *Anaconda Standard,* June 13, 1917, page 7.

2. Lena Sullau quoted from the *Anaconda Standard,* June 13, 1917, page 7. The headline "Sullau Worked to Good Place in Life" appeared above his obituary in the *Butte Miner,* June 10, 1917, page 13.

3. Daniel Harrington, Bureau of Mines/Department of Interior, *Lessons from the Granite Mountain Shaft Fire, Butte (Bureau of Mines Accident Report),* 1922, page 45.

4. Ibid., page 16. In addition to the *Bureau of Mines Accident Report,* the most detailed descriptions of the outbreak of the fire come from newspaper accounts of the inquest by Coroner Aeneas Lane beginning June 20, 1917. See, e.g., *Butte Daily Post,* June 21, 1917, pages 1–2.

5. Gregg S. Clemmer, *American Miners' Carbide Lamps: A Collector's Guide to American Carbide Mine Lighting,* 1997, pages 43–44; *Bureau of Mines Accident Report,* page 38. It was not until spring of 1916 that carbide lanterns had replaced simple candles. See *Bureau of Mines Accident Report,* page 10. On the brightness of early electric lights, see Clemmer, pages 51–52. Information about the miners' dislike for the weight of early electric hand lamps comes from an author interview with Al Hooper conducted on March 5, 2004, at his home in Butte.

6. Author interview with Al Hooper, Butte, Montana. Hooper, ninety-four, was one of the boys who used to earn spending money by salvaging copper wire.

7. *Butte Daily Post,* June 21, 1917, page 2. According to one newspaper account, Baldy Collins had seen the danger and warned Sullau to be careful

with his lamp. It does not appear, however, that Collins repeated this statement in his testimony before the coroner's inquiry. See *Anaconda Standard,* June 10, 1917, page 12.

8. *Bureau of Mines Accident Report,* pages 10, 12.
9. *Butte Daily Post,* June 9, 1917, page 2.
10. Cobb quoted in the *Anaconda Standard,* June 13, 1917, page 12.
11. Ibid.; *Butte Daily Post,* June 21, 1917, page 2.
12. *Anaconda Standard,* June 13, 1917, page 12.
13. The description of the activity on the surface in the immediate aftermath of the fire—particularly the account of the Sheridan/Conroy incident—is drawn largely from Baldy Collins's coroner's inquiry testimony. See *Butte Daily Post,* June 21, 1917, page 2.
14. *Anaconda Standard,* June 10, 1917, page 11.
15. *Bureau of Mines Accident Report,* page 35.
16. *Anaconda Standard,* June 10, 1917, page 11. According to the *Standard,* McLafferty escaped through the High Ore—one of the adjoining mines.
17. *Butte Daily Post,* June 9, 1917, page 2.
18. *Butte Miner,* June 10, 1917, page 1.
19. *Butte Miner,* June 9, 1917, page 1; *Butte Daily Post,* June 9, 1917, page 2.
20. *Butte Miner,* June 10, 1917, pages 1, 12; *Butte Daily Post,* June 9, 1917, pages 1, 2.
21. *Bureau of Mines Accident Report,* page 16.
22. A team from the Montana Bureau of Mines and Geology recently confirmed a claim by the Anaconda Copper Mining Company that Butte today sits atop *10,000 miles* of underground tunnels. *Montana Standard,* June 7, 2004, page 1.
23. *Glossary of Mining Terms,* Bureau of Land Management, Roseburg District Office, www.or.blm.gov.
24. *Butte Miner,* June 10, 1917, page 14. See also *Bureau of Mines Accident Report,* page 42.

Chapter 3. *"The Richest Hill on Earth"*

1. Michael P. Malone, *The Battle for Butte: Mining and Politics on the Northern Frontier, 1864–1906* (1981), page 7; Isaac F. Marcosson, *Anaconda* (1957), page 19.
2. For a description of the diaspora of California miners see Ray Allen Billington, *The Far Western Frontier: 1830–1860* (1956), page 243. For background on placer mining see Michael P. Malone, Richard B. Roeder, and William L.

Lang, *Montana: A History of Two Centuries*, revised edition, 1991, pages 68–69.

3. For descriptions of early mining in Montana, see K. Ross Toole, *Montana: An Uncommon Land*, pages 64-71, and Malone et al., *Montana: A History of Two Centuries*, pages 64–71. Note that Virginia City, Montana, should not be confused with the larger Virginia City, Nevada—site of the famous Comstock silver mine.

4. Biographical information on William A. Clark's early life is drawn from William D. Mangam, *The Clarks, An American Phenomenon* (1941), Chapter 1; Joseph Kinsey Howard, *Montana: High, Wide, and Handsome*, page 58; Marcosson, pages 83–84; Malone, *The Battle for Butte*, pages 12–15; Harry Freeman, *A Brief History of Butte, Montana, 1900*, pages 39–41.

5. Clark is quoted by Marcosson, page 83.

6. Malone, *The Battle for Butte*, page 14.

7. Freeman, page 10; Malone, *The Battle for Butte*, page 7; Marcosson, page 20.

8. Quoted by Malone, *The Battle for Butte*, pages 58–59.

9. Dan Cushman, *Montana—Gold Frontier* (1973), pages 69–70.

10. Quoted by Malone, *The Battle for Butte*, page 59.

11. *The Story of Butte* (1897), available at the Butte-Silver Bow Public Archives, pages 23–24. See also George Everette for the Mai Wah Society, "The Butte Chinese: A Brief History of Chinese Immigrants in Southwest Montana" (undated), available at the Butte-Silver Bow Public Archives, page 5.

12. Everette, page 5.

13. For a succinct summary of the silver period, see *Regional Historic Preservation Plan*, 1993, on file at the Butte-Silver Bow Archives, page II-2. The chapter on Butte history was written by Carrie Johnson.

14. Ibid.; Malone, *The Battle for Butte*, pages 14–15. Clark acquired some of his most significant properties, including the Travona, when his banking clients defaulted on loans. See Malone, *The Battle for Butte*, page 17: "At Butte, as at most mining camps, the discoverers thus lost out to the men of finance."

15. Biographical information about Daly's early life is drawn from Kenneth Ross Toole, *Marcus Daly, A Study of Business in Politics* (1948), master's thesis on file at the University of Montana's Mike and Maureen Mansfield Library in Missoula, pages 1–8; Howard, page 59; Marcosson, pages 2, 41–44; Malone, *The Battle for Butte*, pages 6, 18–20; Freeman, pages 41–45.

16. For discussion of Daly's experiences in Virginia City, see H. Minar Shoebotham, *Anaconda, Life of Marcus Daly, The Copper King* (1956), pages 17–19.

17. Malone, *The Battle for Butte*, page 28.

18. For discussion and analysis of the Alice purchase, see C. B. Glasscock, *The War of the Copper Kings* (1935), pages 45–46; Malone, *The Battle for Butte,* pages 19–20; Malone et al., *Montana: A History of Two Centuries,* page 188; Marcosson, pages 26–27.

19. Malone, *The Battle for Butte,* page 24.

20. See Marcosson, pages 30–31.

21. Hickey's quote is found at UltimateMontana.com under "Anaconda."

22. Hickey dialogue quoted by Marcosson, page 32.

23. Malone, *The Battle for Butte,* page 25. Malone considers and rejects other reports on the formula for Daly's purchase of the Anaconda.

24. Marcosson, page 32.

25. Malone, *The Battle for Butte,* page 27.

26. George Hearst, *The Way It Was* (1890), pages 20–21.

27. The equity shares in the Anaconda were as follows: George Hearst, 39 percent; James Ben Ali Haggin, 26 percent; Marcus Daly, 25 percent; Lloyd Tevis, 10 percent. See Malone, *The Battle for Butte,* page 27.

28. Ibid., pages 27–28. For background on the Hearst-Haggin-Tevis syndicate, see Marcosson, pages 33–36.

29. See Marcosson, page 46, and Malone, *The Battle for Butte,* page 28.

30. Marcosson, pages 48–49.

Chapter 4. "What Men Will Do"

1. *Anaconda Standard,* June 13, 1917, page 7.

2. See *Butte Daily Post,* June 11, 1917, page 8.

3. Daniel Harrington, Bureau of Mines/Department of Interior, *Lessons from the Granite Mountain Shaft Fire, Butte (Bureau of Mines Accident Report),* 1922, page 32.

4. The figure 415 comes from Butte Coroner Aeneas Lane, cited by the *Butte Miner,* June 9, 1917, page 1. The figure 57 comes from the *Bureau of Mines Accident Report,* page 17. There were no active workings between the 1,000- and 1,600-foot levels of either mine.

5. At least six crosscuts connected the Granite Mountain and the Speculator shafts—at the 1,800; 2,000; 2,400; 2,600; 2,800; and 3,000.

6. *Bureau of Mines Accident Report,* page 16.

7. Ibid., page 14.

8. *Butte Daily Post,* June 9, 1917, page 3.

9. Ibid., page 1.

10. *Anaconda Standard,* June 10, 1917, page 11. At least one Speculator cage remained operational and saw some use in raising and lowering rescuers.

11. Author interview with Eddy Drabant (a former Butte miner), April 12, 2004.

12. *Anaconda Standard,* June 11, 1917, page 8.

13. *Butte Miner,* June 10, 1917, page 12.

14. *Bureau of Mines Accident Report,* page 47.

15. *Anaconda Standard,* June 10, 1917, page 1.

16. Author interview with Al Hooper, March 5, 2004. Problems with the Speculator shaft contributed to the North Butte Mining Company's decision to acquire the Granite Mountain property.

17. *Anaconda Standard,* June 10, 1917, page 7.

18. *Bureau of Mines Accident Report,* page 43.

19. Ibid., page 34.

20. Ibid., pages 33–34. It should be noted that at least one contemporaneous news account does state that the mines *had* placed signs above some exits prior to the fire, but that "men, when confused and fearful that delay of a moment will cost them their lives, did not in all cases stop to read the warnings that might have carried some to safety." See the *Anaconda Standard,* June 10, 1917, page 7.

21. *Revised Codes of Montana, 1907,* Section 8541.

22. Quoted by Andrea McCormick, "Men Struggle for Life in Gas-Filled Mines," *Montana Standard,* June 1, 1980, page 23.

23. *Anaconda Standard,* June 11, 1917, page 8.

24. The account of the escape efforts of Jovitich, Bronson, and the men with them is from the *Anaconda Standard,* June 11, 1917, page 8.

25. The account of Wynder and the other men on the 2,200 level is from the *Anaconda Standard,* June 11, 1917, page 8.

26. According to the account in the *Anaconda Standard,* Wynder's partner was a man named Kelly Roberts. Voko, the Finlander, would later tell Wynder that he felt Roberts crawl over him in the darkness—headed in the direction of Granite Mountain.

27. *Bureau of Mines Accident Report,* page 42.

28. *Butte Miner,* June 10, 1917, page 14.

29. *Anaconda Standard,* June 12, 1917, page 1.

30. Ibid.

31. *Butte Daily Post,* June 9, 1917, page 1; *Anaconda Standard,* June 11, 1917, page 8.

32. *Bureau of Mines Accident Report,* page 17.

33. *Anaconda Standard,* June 11, 1917, page 8.

34. *Butte Daily Post,* June 9, 1917, page 1.

35. Ibid., page 3. *Anaconda Standard,* June 11, 1917, page 8.
36. Ibid. Vukovich and four other men from "Balkan states" were honored by their countrymen with a massive funeral cortege through downtown Butte, including a Greek Orthodox priest and 300 marchers clad in the matching "plumes of a marching society." *Anaconda Standard,* June 12, 1917, page 1.
37. *Butte Daily Post,* June 9, 1917, page 1.
38. *Bureau of Mines Accident Report,* page 17. In later days, however, more survivors would be brought up through the Speculator shaft.
39. Ibid., page 30.
40. Ibid.
41. *Anaconda Standard,* June 10, 1917, page 11.
42. Ibid.
43. *Bureau of Mines Accident Report,* page 30.
44. Ibid., pages 30–31.
45. *Anaconda Standard,* June 10, 1917, page 11.
46. "Typical of the miner who is never separated from his 'nosebag,' the bodies of a score of men were found beside their buckets." *Anaconda Standard,* June 11, 1917, page 8. See also the *Butte Daily Post,* June 9, 1917, page 3.
47. *Bureau of Mines Accident Report,* page 31.
48. *Butte Miner,* June 10, 1917, page 14.
49. *Anaconda Standard,* June 10, 1917, pages 11–12.
50. *Butte Daily Post,* June 9, 1917, page 2. For photos and a description of the pulmotor and its use on gas victims, see www.bium.univ-paris5.fr.
51. *Anaconda Standard,* June 10, 1917, page 12.
52. *Butte Daily Post,* June 9, 1917, page 3.
53. *Anaconda Standard,* June 13, 1917, page 7.
54. Burton K. Wheeler, *Yankee from the West* (1962), page 137.
55. *Anaconda Standard,* June 10, 1917, page 12.
56. Ibid., page 11.

Chapter 5. *"Sweetened Corruption"*

1. Descriptions of Daly's wealth are drawn from Isaac F. Marcosson, *Anaconda* (1957), pages 53–54, 62–63; Michael P. Malone, *The Battle for Butte: Mining and Politics on the Northern Frontier, 1864–1906* (1981), page 81.
2. The information on Clark's Fifth Avenue and Santa Barbara residences comes from William D. Mangam, *The Clarks, An American Phenomenon* (1941), pages 83–88; and C. B. Glasscock, *The War of the Copper Kings* (1935), page 287. The information on William Junior's garage came from Mark Reavis,

Butte-Silver Bow Historic Preservation Officer and the house's current owner and resident, interviewed by the author on May 25 and December 16, 2004.

3. Clark quoted by Malone, *The Battle for Butte,* page 85.

4. Malone, *The Battle for Butte,* pages 85–87.

5. The classic study of the Clark-Daly feud is C. B. Glasscock's *The War of the Copper Kings* (1935).

6. Quoted by Malone, *The Battle for Butte,* page 113.

7. Malone, *The Battle for Butte,* pages 123–125.

8. Ibid., page 120.

9. Christopher P. Connolly, *The Devil Learns to Vote* (1938), page 126.

10. Twain cited by Malone, *The Battle for Butte,* page 199.

11. Connolly, pages 138–147.

12. Malone, *The Battle for Butte,* pages 116, 118.

13. Quoted by Connolly, pages 158–159.

14. Malone, *The Battle for Butte,* pages 121–124.

15. Mangam, pages 74–75.

16. Ibid., pages 75–77.

17. Senator Daniel Patrick Moynihan coined this phrase in a different context, but it fits well in turn-of-the-century Montana.

18. Malone, *The Battle for Butte,* page 131.

Chapter 6. "Helmet Men Braving Death"

1. Much of the information about Con O'Neill and his wife, Julia, comes from author interviews (May 17, 2004) with two of their granddaughters—Sally Wolahan Rasmussen and Jeanne Weber. Both Rasmussen and Weber still live in Butte.

2. This account of O'Neill's wife's pleas comes from O'Neill's daughter, Lillian, who gave an interview to the *Montana Standard* printed June 8, 1995— barely two weeks before she died. The description of O'Neill as a "big, robust Irishman" comes from the *Anaconda Standard,* June 10, 1917, page 11.

3. See *Anaconda Standard,* June 10, 1917, page 11. According to the official accident report, gas penetrated the Diamond mine from the Speculator by 1:00 A.M. See Daniel Harrington, Bureau of Mines/Department of Interior, *Lessons from the Granite Mountain Shaft Fire, Butte (Bureau of Mines Accident Report),* 1922, page 16.

4. *Anaconda Standard,* June 10, 1917, page 11.

5. Account drawn from the *Anaconda Standard,* June 10, 1917, page 11, and the *Butte Daily Post,* June 9, 1917, page 2.

6. *Anaconda Standard,* June 10, 1917, page 11.
7. *Butte Daily Post,* June 9, 1917, pages 2–3; the photo of O'Neill appeared in the *Anaconda Standard,* June 10, 1917, page 11.
8. *Bureau of Mines Accident Report,* page 19.
9. Yandell Henderson and James W. Paul, Bureau of Mines/Department of Interior (*Oxygen Apparatus Report*), 1917, page 13. Technical descriptions of the breathing apparatus are also drawn from the *Oxygen Apparatus Report.*
10. *Oxygen Apparatus Report,* page 14. See also the report's discussion on pages 11–14.
11. *Bureau of Mines Accident Report,* page 22.
12. Ibid., page 18.
13. *Butte Daily Post,* June 9, 1917, page 3. See also the *Anaconda Standard,* June 10, 1917, page 7.
14. Ibid.
15. *Anaconda Standard,* June 10, 1917, page 11.
16. Ibid.
17. *Anaconda Standard,* June 10, 1917, page 12.
18. The first car, from Red Lodge, Montana, arrived in Butte around 3:00 P.M. on Saturday, June 9. The second car, from Colorado Springs, Colorado, did not arrive until the morning of June 11, two and a half days after the start of the fire. *Bureau of Mines Accident Report,* page 19.
19. Information on the helmet men is drawn from the *Bureau of Mines Accident Report,* pages 22–23.
20. *Oxygen Apparatus Report,* page 39.
21. *Butte Miner,* June 9, 1917, page 1; *Butte Daily Post,* June 11, 1917, page 1. Among those the soldiers turned back, according to an *Anaconda Standard* account (June 10, 1917, page 7), were "moving picture people." Intriguingly, the paper reported that "a part of the film was completed."
22. *Anaconda Standard,* June 10, 1917, page 11. In its June 9, 1917, edition, page 1, the *Anaconda Standard* reported that a "great volume of black smoke is pouring from the new shaft of the Speculator, filling the ravines below with a black cloud."
23. Except where otherwise indicated, descriptions of the efforts to fight the fire and smoke are drawn from the *Bureau of Mines Accident Report,* pages 17–18.
24. *Anaconda Standard,* June 11, 1917, page 11.
25. *Anaconda Standard,* June 10, 1917, page 7.
26. Though the overall effect of reversing the fans may have been positive, there is at least one report of a negative consequence. Rescuers in the High Ore shaft believed that the reversal of the fans resulted in more fumes being pushed in the direction of the High Ore, "making the recovery of bodies

through this route impossible." Ultimately, bulkheads were thrown up to protect the High Ore workings from the North Butte fumes. *Butte Miner,* June 10, 1917, page 1.

27. See the *Bureau of Mines Accident Report,* page 23.

28. *Anaconda Standard,* June 11, 1917, page 11; *Butte Daily Post,* June 11, 1917, page 8. The *Anaconda Standard,* June 14, 1917, page 7, reported that the fire "is confined to small pockets on the station, and the blaze is absolutely under control."

29. According to the *Butte Miner* (June 10, 1917, page 11), at least ten stations caved in. The entire shaft would ultimately have to be retimbered.

30. *Anaconda Standard,* June 11, 1917, page 11.

31. *Anaconda Standard,* June 10, 1917, page 12.

32. *Butte Miner,* June 9, 1917, page 1. According to the *Bureau of Mines Accident Report,* page 22, "at least 15 previously untrained men" wore apparatus in order to serve as guides.

33. *Bureau of Mines Accident Report,* page 20.

34. See, e.g., *Bureau of Mines Accident Report,* page 22.

35. *Anaconda Standard,* June 11, 1917, page 11.

36. *Oxygen Apparatus Report,* page 61.

37. *Anaconda Standard,* June 10, 1917, page 12.

38. *Bureau of Mines Accident Report,* page 23.

39. *Anaconda Standard,* June 12, 1917, page 1.

40. *Butte Daily Post,* June 9, 1917, page 2. The article misspells Budelière's name as "Budilier."

41. The account of Budelière is drawn from the *Butte Miner,* June 9, 1917, page 15.

42. *Bureau of Mines Accident Report,* page 23.

43. Descriptions of the triage are drawn primarily from an account in the *Anaconda Standard,* June 9, 1917, page 11. See also *Anaconda Standard,* June 10, 1917, page 3.

44. *Butte Miner,* June 9, 1917, page 1.

45. *Anaconda Standard,* June 9, 1917, page 11.

46. *Butte Daily Post,* June 9, 1917, page 1.

Chapter 7. "Standard Oil Coffins"

1. Technically, the corporate entity of Standard Oil Company was separated from the purchase of Anaconda, because Anaconda and its various parts were folded into a newly created holding company—"Amalgamated." This distinc-

tion, however, is more form than substance. The Anaconda transaction was consummated by Standard Oil principals William G. Rockefeller and Henry Rogers. Until the 1915 liquidation of Amalgamated and reestablishment of Anaconda as a freestanding company, "Amalgamated" and "Anaconda" were effectively synonymous. Contemporaneous accounts often refer to Amalgamated management as "Standard Oil."

2. *New York Times,* April 28, 1899, page 1.

3. See Isaac F. Marcosson, *Anaconda* (1957), pages 54–56; K. Ross Toole, *20th Century Montana: A State of Extremes* (1972), pages 101–104.

4. K. Ross Toole, *Montana: An Uncommon Land* (1959), page 195.

5. A classic account of the robber baron era is Matthew Josephson's *The Robber Barons* (1934). See pages 381–389 for a general discussion, and pages 397–398 for a specific discussion of Standard Oil and Anaconda.

6. Rogers quoted by C. B. Glasscock in his classic study of Butte, *The War of the Copper Kings* (1935), page 187.

7. The best description of Standard Oil's takeover of Anaconda is found in Thomas W. Lawson's *Frenzied Finance* (1905). Lawson, a Boston financier who worked with Rogers and Rockefeller in molding the deal, later turned on his former partners and published his "tell all"–style book, generally thought to be a credible account.

8. Glasscock, page 197.

9. See Lawson, pages 223–225, and Josephson, page 398.

10. Biographical information about Heinze comes primarily from Sarah McNelis, *Copper King at War: The Biography of F. Augustus Heinze* (1968), pages 2–5, 9, 13.

11. Reno H. Sales, *Underground Warfare at Butte* (1964), pages 8–10.

12. McNelis, page 18.

13. Glasscock, page 124.

14. Michael P. Malone, *The Battle for Butte: Mining and Politics on the Northern Frontier, 1864–1906* (1981), page 168.

15. Malone, *The Battle for Butte,* pages 147–148. *Anaconda Standard* cited by Malone.

16. Clancy story told by Marcosson, page 120.

17. Clancy quoted by Sales, page 23.

18. Malone, *The Battle for Butte,* pages 148–149.

19. Ibid., page 148.

20. Cited by Malone, *The Battle for Butte,* page 155.

21. See Malone, *The Battle for Butte,* pages 153–154, 164.

22. Ibid., page 153.

23. Ibid., pages 160–161.

24. Ibid., page 157.

25. Sales, pages 16–17.
26. Ibid., page 29.
27. Ibid., pages 30–34. Reno Sales, the author of *Underground Warfare at Butte*, was one of the Amalgamated "spies."
28. Ibid., pages 35–39.
29. Ibid., pages 56–58.
30. The casualties, both employees of Amalgamated, died while trying to construct a bulkhead to block fumes from a fire set in the Rarus. Heinze men, working nearby to cover evidence of their mining and apparently unaware of the two Amalgamated miners, set off a huge explosion. The shock wave blew wood from the bulkhead into the Amalgamated men, killing one instantly and fatally wounding the other. Ibid., pages 42–44.
31. Malone, Roeder, and Lang, *Montana: A History of Two Centuries* (1976), pages 179–181.
32. Malone, *The Battle for Butte*, pages 168, 187.
33. Ibid., page 172.
34. Judge Clancy's ruling was summarized in the Montana Supreme Court decision that ultimately overturned it—*MacGinniss v. Boston & Montana Consolidated Copper & Silver Mining Company et al,* 29 Montana Reporter 433 (decided February 1, 1904).
35. Malone, *The Battle for Butte*, page 173.
36. The description of the aftermath of the Clancy decision, including the text of the Heinze courthouse speech, is drawn primarily from Malone, *The Battle for Butte*, pages 173–179.
37. McNelis, page 83; Malone, *The Battle for Butte*, page 177.
38. Ibid.
39. Burton K. Wheeler, *Yankee from the West* (1962), page 77.
40. Biographical information about Wheeler (including the poker anecdote) is drawn primarily from his autobiography, *Yankee from the West* (1962), pages 57–63.

Chapter 8. *"Then We Met Duggan"*

1. *Anaconda Standard,* June 11, 1917, page 1.
2. Ibid.
3. The incident at the Speculator/High Ore connection is discussed by Daniel Harrington, Bureau of Mines/Department of Interior, *Lessons from the Granite Mountain Shaft Fire, Butte* (*Bureau of Mines Accident Report*), 1922, page 44.

4. Ibid.

5. See *Bureau of Mines Accident Report,* page 44.

6. The legality of doorless bulkheads is discussed at length in Chapter 18.

7. Burton K. Wheeler, *Yankee from the West,* 1962, page 137.

8. Arnon Gutfeld, "The Speculator Disaster," *Arizona and the West: A Quarterly Journal of History,* Spring 1969, page 29.

9. *Anaconda Standard,* June 11, 1917, page 1.

10. *Anaconda Standard,* June 13, 1917, page 12.

11. One newspaper account lists Duggan's birthplace as Highland, Pennsylvania. See *Anaconda Standard,* June 15, 1917, page 7.

12. Most biographical information about Manus Duggan comes from author interviews with his daughter, Manus Dugan Banko, conducted by telephone on March 11, 2004; June 10, 2004; and November 9, 2004. Information about Manus's mother is from the *Butte Miner,* June 10, 1917, page 14.

    There is some question as to the proper spelling of Manus's last name. The Butte newspapers all spelled the name "Duggan"—with two *g*s. Duggan's daughter is in possession of a check endorsed by the miner that he signed "Duggan." All contemporaneous newspaper accounts also spell the name with two *g*s. However, according to Duggan's daughter, Duggan's wife, Madge, later dropped one *g*, spelling her name "Dugan." To further the confusion, Manus Duggan's mother's *maiden name* was Dugan—with one *g*.

13. Ibid.

14. *Montana Standard,* November 5, 1978, page 22.

15. Manus Duggan's house, at 1010 Zarelda Street, is still standing today.

16. Josiah James's accounts are found in the *Anaconda Standard,* June 11, 1917, page 8, and in an article by Andrea McCormick, "Men Struggle for Life in Gas-Filled Mines," *Montana Standard,* June 1, 1980, page 23.

17. *Butte Miner,* June 11, 1917, page 1.

18. *Anaconda Standard,* June 13, 1917, page 7.

19. *Anaconda Standard,* June 12, 1917, page 1.

20. *Anaconda Standard,* June 11, 1917, page 8.

21. See Albert Cobb account, *Anaconda Standard,* June 13, 1917, page 12.

22. Ibid.

23. See *Bureau of Mines Accident Report,* pages 12, 24.

24. *Anaconda Standard,* June 13, 1917, page 12; *Bureau of Mines Accident Report,* page 24. "Lagging" was the miners' term for a plank or board. For an excellent collection of mining lingo and "anti-dotes," see *Memories of a Mining Camp* (1999) by Jim Edwards and Kevin Shannon. The Wirta account is from the *Anaconda Standard,* June 11, 1917, page 1.

25. See *Bureau of Mines Accident Report,* page 25; Josiah James account in the *Anaconda Standard,* June 11, 1917, page 8.

26. *Butte Miner,* June 11, 1917, page 1.

27. Charles Negretto account, *Anaconda Standard,* June 12, 1917, page 1.

28. *Anaconda Standard,* June 11, 1917, pages 8–9. See also the *Bureau of Mines Accident Report,* page 25.

29. *Anaconda Standard,* June 11, 1917, page 11.

30. *Anaconda Standard,* June 11, 1917, page 8; *Bureau of Mines Accident Report,* page 25. The Cobb account is found in the *Anaconda Standard,* June 13, 1917, page 12.

31. Lucas account, *Butte Miner,* June 11, 1917, page 1.

32. Ibid.

33. See *Bureau of Mines Accident Report,* page 25.

34. The physical conditions behind the Duggan bulkhead are described in the *Bureau of Mines Accident Report,* pages 25, 39. The reference to a flashlight comes from the Wirta account, *Anaconda Standard,* June 11, 1917, page 1: "I had my watch, the only one in the crowd, and a flashlight." (In actuality, Charles Negretto also had a watch. See *Anaconda Standard,* June 12, 1917, page 1.)

35. *Anaconda Standard,* June 11, 1917, page 8.

36. Josiah James account, *Anaconda Standard,* June 11, 1919, page 8.

Chapter 9. *"Hey Jack, What the Hell"*

1. *Copper Camp* (1943), page 19. The 1943 *Copper Camp* is the classic treatment on the history, legend, and lore of daily life in Butte. It was written by a man named William A. Burke. Burke, though, wrote under the auspices of the Montana Writers' Project, a Depression-era program whose rules prohibited an individual from being named as the author. A more recent, scholarly approach to Butte's cultural history is Mary Murphy's excellent *Mining Cultures: Men, Women, and Leisure in Butte, 1914–41* (1997).

2. Michael P. Malone, *The Battle for Butte* (1981), page 202.

3. For a discussion of the development of Chilean copper resources see Isaac F. Marcosson, *Anaconda* (1957), Chapter 9.

4. Malone, page 202; Marcosson, page 199.

5. Mary Murphy, *Mining Cultures* (1997), pages 10, 14. According to U.S. Census figures for 1910, the Irish represented 24.2 percent of Butte's foreign-born population, while Slavs, Italians, and Finnish together consti-

tuted 9.3 percent. By 1920, the percentage of foreign-born from Ireland had fallen to 20.7 percent, while Slavs, Italians, and Finns had grown to a collective 20.1 percent. Cited by Jerry W. Calvert, *The Gibraltar: Socialism and Labor in Butte, Montana, 1895–1920* (1988), page 59.

6. Handbill cited by Calvert, page 103.

7. Brian Shovers, "The Perils of Working in the Butte Underground: Industrial Fatalities in the Copper Mines, 1880–1920," *Montana* (Spring 1987), pages 30–31.

8. Shovers, pages 38–39. James D. Harrington, *Mining Related Fatalities in the Butte & Anaconda Regions: A Working Document* (1992, updated March 4, 1997), unpublished data on file at the Butte-Silver Bow Public Archives in Butte, Montana.

9. Ray Stannard Baker, "Butte City: Greatest of Copper Camps," *The Century Magazine* (April 1903), page 875.

10. Clark cited by Michael P. Malone, *The Battle for Butte,* page 89.

11. *Copper Camp* (1943), pages 103–104.

12. Author interview with Jim Keane, October 23, 2004.

13. The best summary of the hazards faced by Butte miners is Brian Shovers's "The Perils of Working in the Butte Underground: Industrial Fatalities in the Copper Mines, 1880–1920," pages 26–39. On the 1915 explosion, see the *Butte Miner,* October 20, 1915, page 1.

14. *Copper Camp,* pages 171–172; author interview with Al Hooper, a former hoist operator, September 4, 2003; Shovers, pages 26, 28.

15. Author interview with Mark Reavis, Director for Historic Preservation, Butte, May 25, 2004.

16. *Copper Camp,* page 170.

17. Bureau of Mines, *Miners' Consumption in the Mines of Butte, Montana: Preliminary Report of an Investigation Made in the Years 1916–1919,* 1921, page 11.

18. Shovers, page 28, citing the U.S. Department of Labor.

19. Robert Wallace, *The Miners* (1976), page 94.

20. James D. Harrington, *Mining Related Fatalities in the Butte & Anaconda Regions: A Working Document,* updated March 4, 1997, on file at the Butte-Silver Bow Public Archives in Butte, Montana.

21. Author interview with Keane.

22. See Murphy, pages 19, 49–50.

23. *Copper Camp,* page 19.

24. *Anaconda Standard,* June 15, 1917, page 11.

25. *Copper Camp,* page 9.

26. Ibid., pages 183–190.

27. *Anaconda Standard,* June 16, 1917, page 4. See also Murphy, page 8.

28. *Anaconda Standard,* June 15, 1917, page 7. Butte's largest "photoplay" house was the Rialto. On June 15, 1917, the Rialto was showing a "U.S. government film" called *The Peacemakers,* featuring footage of "U.S. Troops Under Fire." *Butte Daily Post,* June 15, 1917, page 5.

29. The information on libraries comes from the *1917 Butte Buyers' Guide,* a sort of city directory, available at the Butte-Silver Bow Public Archives.

30. Ibid.

31. Author interview with Manus Dugan Banko, June 10, 2004.

32. Murphy, page 114.

33. Ibid., pages 114–118.

34. *Copper Camp,* pages 239–242.

35. *Butte Miner,* June 10, 1917, page 12. Boxing information on Leo Bens comes from the *Vermont Boxing History & International Pugilist Review,* http://esf .uvm.edu/btbox, citing the *Brattleboro Daily Reformer,* January 10, 1916, page 4.

36. *Butte Daily Post,* June 21, 1917, page 10; *Anaconda Standard,* September 5, 1917, page 11.

37. An excellent analysis of the role of Catholicism in Butte can be found in Chapter 4 ("Church, Party, and Fraternity: The Irish and Their Associations") of David M. Eamons's *The Butte Irish: Class and Ethnicity in an American Mining Town, 1875–1925.* Statistics on churches and synagogues are from the *1917 Butte Buyers' Guide.*

38. *Butte Miner,* June 10, 1917, page 14.

39. Ibid., page 12.

40. See, e.g., *Butte Miner,* June 10, 1917, page 15.

41. *Butte Miner,* June 10, 1917, page 12. See also *Butte Daily Post,* June 12, 1917, page 2.

42. *Butte Daily Post,* June 11, 1917, page 3.

43. *Butte Miner,* June 12, 1917, page 6.

44. *Anaconda Standard,* June 11, 1917, page 7.

45. *Butte Daily Post,* June 9, 1917, page 3.

46. *Anaconda Standard,* June 11, 1917, page 8.

Chapter 10. "If the Worst Comes"

1. Duggan's note was first published in the *Butte Daily Post,* June 14, 1917, page 1.

2. While there are only two Duggan notes, it appears likely that the first note was written in *two separate installments,* probably separated in time by about

twelve hours. The first note is headed "Saturday morning, 8:45." The first paragraph ends with Duggan describing the men as "[a]ll in good spirits." The second paragraph opens with the line "I realize that all oxygen has just been consumed." From the accounts of numerous men behind the bulkhead, it does not appear that oxygen depletion was an issue until around *8:00 on Saturday night*. The notion that the first note was written in two installments is also supported by the dramatic shift in tone between the first and second paragraphs.

3. Daniel Harrington, Bureau of Mines/Department of Interior, *Lessons from the Granite Mountain Shaft Fire, Butte (Bureau of Mines Accident Report)*, 1922, page 25. Martin Novak account, *Anaconda Standard*, June 12, 1917, page 11.

4. Shea account, *Anaconda Standard*, June 11, 1917, page 8.

5. Still account, *Anaconda Standard*, June 11, 1917, page 11.

6. Mary Murphy, *Mining Cultures* (1997), page 16.

7. Cobb account, *Anaconda Standard*, June 13, 1917, page 12.

8. *Anaconda Standard*, June 11, 1917, page 8.

9. *Bureau of Mines Accident Report*, page 25.

10. Ibid., page 30.

11. Daniel Harrington does the actual math on the air supply (based on the 1917-era knowledge of oxygen consumption rates) in the *Bureau of Mines Accident Report*, page 26.

12. Wirta account, *Anaconda Standard*, June 11, 1917, page 11.

13. Lucas account, *Butte Miner*, June 11, 1917, page 1.

14. Wirta account, *Anaconda Standard*, June 11, 1917, page 11.

15. Ibid.

16. Shea account, *Anaconda Standard*, June 11, 1917, page 8.

17. Galia account, *Anaconda Standard*, June 13, 1917, page 7.

18. Still account, *Anaconda Standard*, June 11, 1917, page 11. See also *Bureau of Mines Accident Report*, page 25.

19. Novak account, *Anaconda Standard*, June 12, 1917, page 11.

20. Still account, *Anaconda Standard*, June 11, 1917, page 11.

21. Wirta account, *Anaconda Standard*, June 11, 1917, page 1.

22. See, e.g., Wirta account, *Anaconda Standard*, June 11, page 9.

23. Ibid., page 11.

24. Cobb account, *Anaconda Standard*, June 13, 1917, page 12. See also *Bureau of Mines Accident Report*, page 26.

25. The account of the "alien's" prayer comes from the *Anaconda Standard*, June 11, 1917, page 8.

Chapter 11. "Dreading to Look"

1. *Anaconda Standard,* June 12, 1917, page 1.
2. Author interview with Manus Dugan Banko, November 11, 2004.
3. *Butte Miner,* June 9, 1917, page 1.
4. *Anaconda Standard,* June 9, 1917, page 1; *Anaconda Standard,* June 10, 1917, page 11. The *Anaconda Standard* published a self-congratulatory article in its June 10 edition bragging about the "57 minute" run made by its delivery truck between Anaconda (the town) and Butte—a distance of twenty-three miles. At times, noted the article, the truck did "better than 40 miles an hour." *Anaconda Standard,* June 10, 1917, page 12.
5. *Butte Miner,* June 11, 1917, page 5; *Anaconda Standard,* June 11, 1917, page 11.
6. *Anaconda Standard,* June 10, 1917, page 11.
7. Ibid.
8. Ibid. The list of the dead was prepared by Butte historian James D. Harrington, *Mining Related Fatalities in the Butte & Anaconda Regions: A Working Document,* updated March 4, 1997, on file at the Butte-Silver Bow Public Archives in Butte, Montana.
9. *Anaconda Standard,* June 10, 1917, page 11.
10. *Butte Daily Post,* June 9, 1917, page 1.
11. *Washington Post,* June 10, 1917, page 1. See also *New York Times,* June 10, 1917, page 8.
12. Letter from Margaret O'Neil to North Butte Officials, June 13, 1917, on file at the Montana Historical Society Archives, Helena.
13. *Anaconda Standard,* June 14, 1917, page 9; *Anaconda Standard,* June 11, 1917, page 1.
14. B. Thompson correspondence with North Butte officials on file at the Montana Historical Society Archives, Helena. The archives has a thick folder of the letters and telegraphs from relatives of the North Butte miners, and the pain that fills them is still palpable today.
15. Letter from Mollie J. Powers to the "Superintendant of the North Butte Mine," June 18, 1914, on file at the Montana Historical Society Archives, Helena.
16. Letter from Mrs. K. Douglas of Denver Colorado to the "Supt. Of Granite Mountain Mine," June 18, 1917, on file at the Montana Historical Society Archives, Helena.
17. *Anaconda Standard,* June 11, 1917, page 9.
18. *Butte Daily Post,* June 9, 1917, page 1.

19. *Anaconda Standard,* June 10, 1917, page 11; *Anaconda Standard,* June 11, 1917, page 8.

20. *Butte Miner,* June 11, 1917, page 5.

21. *Anaconda Standard,* June 11, 1917, page 8.

22. *Anaconda Standard,* June 12, 1917, page 11.

23. *Butte Miner,* June 11, 1917, page 1.

24. *Anaconda Standard,* June 11, 1917, page 9.

25. *Anaconda Standard,* June 11, 1917, page 8.

26. *Butte Daily Post,* June 11, 1917, page 1.

27. *Butte Miner,* June 11, 1917, page 1.

28. *Anaconda Standard,* June 12, 1917, pages 1, 11.

29. *Daily Missoulian,* June 12, 1917, page 1; *Anaconda Standard,* June 12, 1917, page 7. On June 13, the *Anaconda Standard* offered an editorial thank-you to its sister city: "The kindest hearted, most thoughtful, most sympathetic people on earth live in Missoula . . . There are flowers everywhere. They do much to relieve the tenseness and somberness of Butte's greatest grief and brighten these days of utter sadness." *Anaconda Standard,* June 13, 1917, page 6.

30. *Anaconda Standard,* June 10, 1917, page 1.

31. Ibid.

32. Ibid., page 12.

33. *Anaconda Standard,* June 11, 1917, page 5; *Butte Miner,* June 10, 1917, page 14.

34. Letter from C. M. Covert to the North Butte Mining Company, June 18, 1917. Letter on file at the Montana Historical Society Archives, Helena. In response to the letter, the North Butte sent back a terse response, "[W]e are unable to give you any advice in the matter."

35. *Butte Miner,* June 10, 1917, page 6; *Butte Miner,* June 11, 1917, page 5.

36. The Montana Compensation Law is discussed in detail in Chapter 23, "What Will Become of Them."

37. *Anaconda Standard,* June 10, 1917, page 11.

38. *Butte Daily Post,* June 11, 1917, page 8; *Butte Miner,* June 10, 1917, page 11.

39. *Butte Miner,* June 11, 1917, page 5.

Chapter 12. "Now Is the Time"

1. The dialogue between Cobb and Duggan is quoted from Cobb's account in the *Anaconda Standard,* June 13, 1917, page 12.

2. *Butte Miner,* June 11, 1917, page 1.

3. La Montague account, *Butte Miner,* June 11, 1917, page 8. See also photo of La Montague and its caption in the *Anaconda Standard,* June 11, 1917, page 1.

4. According to Wirta, "I had my watch, the only one in the crowd, and a flashlight." *Anaconda Standard,* June 11, 1917, page 9.

5. The analysis of the air in the Duggan bulkhead is drawn from Daniel Harrington, Bureau of Mines/Department of Interior, *Lessons from the Granite Mountain Shaft Fire, Butte (Bureau of Mines Accident Report),* 1922, page 26. The author asked Dr. Brent Ruby, Director of the Human Performance Laboratory at the University of Montana, to review the conclusions of the *Bureau of Mines Accident Report.* While it is not possible to replicate the conditions behind the bulkhead, the general facts applied in the report (e.g., the oxygen content in "normal" air) were appropriate.

6. See *Anaconda Standard,* June 11, page 8; *Bureau of Mines Accident Report,* page 25; Lucas account, *Butte Miner,* June 11, 1917, page 1: "[t]he gas would creep in a little bit at a time."

7. *Bureau of Mines Accident Report,* page 26.

8. *Anaconda Standard,* June 11, 1917, page 8.

9. Ibid.

10. Quoted by Josiah James in his account in the *Anaconda Standard,* June 13, 1917, page 12.

11. James account, *Anaconda Standard,* June 13, 1917, page 12.

12. Lucas account, *Butte Miner,* June 11, 1917, page 1.

13. La Montague account, *Butte Miner,* June 11, 1917, page 8.

14. See *Anaconda Standard,* June 12, 1917, page 1.

15. James account, *Anaconda Standard,* June 11, 1917, page 8.

16. Novak account, *Anaconda Standard,* June 12, 1917, page 11.

17. James account, *Anaconda Standard,* June 11, 1917, page 8.

18. Quote from Charles Negretto, *Anaconda Standard,* June 12, 1917, page 1.

19. Photo caption, *Anaconda Standard,* June 11, 1917, page 1.

20. *Anaconda Standard,* June 11, 1917, page 8.

21. Wirta account, *Anaconda Standard,* June 11, 1917, page 1.

22. Galia account, *Anaconda Standard,* June 13, 1917, page 7.

23. Wirta account, *Anaconda Standard,* June 11, 1917, page 1.

24. *Anaconda Standard,* June 12, 1917, page 1. This article spells Heston's name "Hastings."

25. Wirta account, *Anaconda Standard,* June 11, 1917, page 1.

26. Duggan is quoted by Josiah James in his account in the *Anaconda Standard,* June 11, 1917, page 8.

Chapter 13. *"Men Alive!"*

1. *Butte Daily Post,* June 11, 1917, page 1.
2. Wirta account, *Anaconda Standard,* June 11, 1917, page 1.
3. James account, *Anaconda Standard,* June 11, 1917, page 8.
4. Lucas account, *Butte Miner,* June 11, 1917, page 1.
5. According to one account, unconfirmed by any other document, helmet men actually reached the Duggan bulkhead—even touched it—at 3:00 P.M. on Saturday afternoon. Because of the thick smoke present, however, they were unable to discern the structure's significance. Nor would they have been able to read the words "men in here" apparently written on the bulkhead exterior by Albert Cobb. See *Anaconda Standard,* June 11, 1917, page 8.
6. James account, *Anaconda Standard,* June 11, 1917, page 8.
7. *Anaconda Standard,* June 11, 1917, page 1.
8. Cobb account, *Anaconda Standard,* June 13, 1917, page 12.
9. Daniel Harrington, Bureau of Mines/Department of Interior, *Lessons from the Granite Mountain Shaft Fire, Butte (Bureau of Mines Accident Report),* 1922, page 26.
10. James account, *Anaconda Standard,* June 11, 1917, page 8.
11. Reverend Wilson's sermon was printed in the *Anaconda Standard,* June 11, 1917, page 7. The figure of forty-four churches is drawn from the 1917 *Butte Buyers' Guide,* the functional equivalent of today's Yellow Pages.
12. Wirta account, *Anaconda Standard,* June 11, 1917, page 1.
13. See James account in an article by Andrea McCormick, "Men Struggle for Life in Gas-Filled Mines," *Montana Standard,* June 1, 1980, page 23.
14. Ibid.
15. *Anaconda Standard,* June 11, 1917, page 8.
16. *Butte Daily Post,* June 11, 1917, page 1.
17. *Butte Miner,* June 11, 1917, page 1; *Anaconda Standard,* June 11, 1917, page 11.
18. *Butte Miner,* June 11, 1917, page 1.
19. *Butte Daily Post,* June 11, 1917, page 1; *Anaconda Standard,* June 11, 1917, page 8.
20. *Anaconda Standard,* June 11, 1917, page 11; *Butte Daily Post,* June 11, 1917, page 1.
21. The entire passage on Cobb is drawn from his account in the *Anaconda Standard,* June 13, 1917, page 12.
22. *Anaconda Standard,* June 11, 1917, page 8.
23. Ibid.
24. Ibid.

25. *Butte Miner,* June 11, 1917, pages 1, 8.
26. See *Butte Daily Post,* June 11, 1917, page 1.
27. *Anaconda Standard,* June 11, 1917, page 8.
28. *Butte Miner,* June 11, 1917, page 8.
29. The account of Josiah James appears in an article by Andrea McCormick, "Men Struggle for Life in Gas-Filled Mines," *Montana Standard,* June 1, 1980, page 23.
30. Josiah James's note is reprinted in the McCormick article.
31. The Ledvina/Galia incident and dialogue were reported in the *Butte Daily Post,* June 13, 1917, page 7.
32. *Anaconda Standard,* June 13, 1917, page 7.
33. *Anaconda Standard,* June 11, 1917, page 1.
34. *Butte Miner,* June 11, 1917, page 1.
35. *Bureau of Mines Accident Report,* page 26.
36. Bodnaruk account, *Anaconda Standard,* June 11, 1917, page 1.
37. *Anaconda Standard,* June 11, 1917, page 1.
38. Ibid.
39. *Anaconda Standard,* June 13, 1917, page 12.

Chapter 14. "Bamboozeling or Abuse"

1. The complete text of the strikers' leaflet was printed in the State of Montana's *First Biennial Report of the Department of Labor and Industry, 1913–1914,* pages 30–31.
2. Interview of Burton K. Wheeler by John Paxon, February 7, 1972, page 2. Transcript on file at the Montana Historical Society Archives, Helena.
3. Wheeler, page 80.
4. Ibid., page 90.
5. Ibid., page 95.
6. Michael P. Malone, Richard B. Roeder, and William L. Lang, *Montana: A History of Two Centuries* (1976), pages 254–258.
7. Wheeler, pages 109–111.
8. *Butte Daily Post,* June 11, 1917, page 8.
9. Jerry W. Calvert, *The Gibraltar: Socialism and Labor in Butte, Montana, 1895–1920* (1988), page 71; Paul F. Brissenden, "The Butte Miners and the Rustling Card," *The American Economic Review,* vol. 10, no. 4 (December 1920), page 756.
10. Calvert, page 4; Wheeler, page 89.
11. Malone, Roeder, and Lang, pages 207, 271; Michael P. Malone, *The Battle*

for Butte: Mining and Politics on the Northern Frontier, 1864–1906 (1995), page 151.

12. Malone, *The Battle for Butte,* page 203.

13. Brian Shovers, "The Perils of Working in the Butte Underground: Industrial Fatalities in the Copper Mines, 1880–1920," *Montana* (Spring 1987), pages 33–34. *First Biennial Report,* page 17.

14. Calvert, pages 24–25, 43, 76.

15. Paul F. Brissenden, "The Butte Miners and the Rustling Card," *The American Economic Review,* vol. 10, no. 4 (December 1920), pages 761–764.

16. Calvert, pages 78–79.

17. George R. Tompkins, *The Truth About Butte* (1917), page 20.

18. Brissenden, page 772.

19. Ibid., page 756.

20. *Industrial Worker,* January 13, 1913, cited by Calvert, page 163.

21. These contracts, it is important to note, were endorsed by a majority of the members of the Butte Miners' Union, though there is some indication—at least in 1907—that union leaders did not adequately explain the working of the sliding scale system. See Calvert, page 72.

22. *First Biennial Report,* page 25; Calvert, pages 72, 76–78.

23. Calvert, page 77.

24. *First Biennial Report,* page 25.

25. Calvert, pages 81–82.

26. Ibid., pages 82–83; *First Biennial Report,* pages 25–26.

27. Description of the events of the night of June 23, 1914, are drawn from the *First Biennial Report,* pages 27–28.

28. George Tompkins, *The Truth About Butte* (1917), page 30.

29. *First Biennial Report,* pages 29–31; Calvert, page 85.

30. *First Biennial Report,* pages 30–31.

31. Ibid., page 31; Calvert, pages 87–88.

32. *First Biennial Report,* pages 32–33; Calvert, pages 87–88.

33. *First Biennial Report,* pages 33–34.

Chapter 15. *"Too Good Miners"*

1. Marthey account, *Anaconda Standard,* June 12, 1917, page 11.

2. Descriptions of La Martine's discovery of a bulkhead at the 2,200 are drawn from the *Anaconda Standard,* June 12, 1917, page 11, and the *Butte Daily Post,* June 11, 1917, page 1.

3. *Anaconda Standard,* June 12, 1917, page 11. Emphasis added.

4. *Anaconda Standard,* June 12, 1917, page 7.

5. *Butte Daily Post,* June 11, 1917, page 1.

6. *Anaconda Standard,* June 12, 1917, page 7.

7. E. A. Osborne, quoted in the *Anaconda Standard,* June 13, 1917, page 9.

8. Marthey account, *Anaconda Standard,* June 12, 1917, page 11.

9. Garrity account, *Butte Daily Post,* June 11, 1917, page 7.

10. Marthey account, *Anaconda Standard,* June 12, 1917, page 11.

11. *Anaconda Standard,* June 13, 1917, page 9.

12. Garrity account, *Butte Daily Post,* June 11, 1917, page 7.

13. Discussion of the air hose is from Daniel Harrington, Bureau of Mines/Department of Interior, *Lessons from the Granite Mountain Shaft Fire, Butte (Bureau of Mines Accident Report),* 1922, page 27.

14. *Bureau of Mines Accident Report,* page 27.

15. Ibid.

16. *Anaconda Standard,* June 12, 1917, page 11.

17. *Bureau of Mines Accident Report,* page 27.

18. Marthey account, *Anaconda Standard,* June 12, 1917, page 11.

19. Contemporaneous accounts refer consistently to ten men in the Moore party. However, the *Bureau of Mines Accident Report* states that there were *eight* men in the Moore party, without ever explaining the discrepancy with contemporaneous accounts. See Daniel Harrington, Bureau of Mines/Department of Interior, *Lessons from the Granite Mountain Shaft Fire, Butte (Bureau of Mines Accident Report),* 1922, page 28. For further discussion on this point, see Chapter 17: "We're Dying in Here."

20. *Anaconda Standard,* June 12, 1917, page 1.

21. Marthey account, *Anaconda Standard,* June 12, 1917, page 11.

22. Garrity account, *Anaconda Standard,* June 12, 1917, page 1.

23. See *Bureau of Mines Report,* page 29. The author asked Dr. Brent Ruby, Director of the Human Performance Laboratory at the University of Montana, to review the conclusions of the *Bureau of Mines Accident Report.* While it is not possible to replicate the conditions behind the bulkhead, the general facts applied in the report (e.g., the oxygen content in "normal" air) were appropriate.

24. *Anaconda Standard,* June 11, 1917, page 11. *Bureau of Mines Accident Report,* 28.

25. Marthey account, *Anaconda Standard,* June 12, 1917, page 11.

26. Moore's letters were published by the *Butte Miner,* June 14, 1917, page 8.

27. Marthey account, *Anaconda Standard,* June 12, 1917, page 11: "He said he had $5,000 life insurance, $3,000 in the bank and other moneys . . ."

28. The ellipses were inserted by the *Butte Miner*, "the deleted portions being of a personal nature between Mr. Moore and his wife." *Butte Miner*, June 14, 1917, page 8.
29. Ibid.

Chapter 16. "In the Dark"

1. Moore's letters were published by the *Butte Miner*, June 14, 1917, page 8.
2. See Daniel Harrington, Bureau of Mines/Department of Interior, *Lessons from the Granite Mountain Shaft Fire, Butte (Bureau of Mines Accident Report)*, 1922, page 28. According to the report, "When Moore's candle burned down, one carbide lamp was kept burning for about an hour, and then the entire party was in darkness."
3. Author interview with Eddy Drabendt, former Butte miner, April 12, 2004.
4. *Bureau of Mines Accident Report*, page 28.
5. The reasons contributing to the poor air behind the Moore bulkhead are analyzed in the *Bureau of Mines Accident Report*, pages 28–30.
6. *Bureau of Mines Accident Report*, page 29.
7. According to the *Bureau of Mines Accident Report*, the Moore chamber was approximately 130 feet long by seven feet wide by seven feet high—giving 6,500 cubic feet of air. Ibid., page 39.
8. Garrity account, *Anaconda Standard*, June 12, 1917, page 1.
9. Marthey account, *Anaconda Standard*, June 12, 1917, page 11.
10. Ibid., pages 11, 7.
11. *Butte Daily Post*, June 11, 1917, page 1.
12. Marthey account, *Anaconda Standard*, June 12, 1917, page 11.
13. *Anaconda Standard*, June 14, 1917, page 7.
14. Marthey account, *Anaconda Standard*, June 12, 1917, page 11.
15. *Bureau of Mines Accident Report*, page 29.
16. *Anaconda Standard*, June 12, 1917, page 11.
17. *Butte Miner*, June 11, 1917, page 1; *Anaconda Standard*, June 11, 1917, page 1.
18. *Anaconda Standard*, June 12, 1917, page 1, 11; *Anaconda Standard*, June 11, 1917, page 9.
19. *Butte Miner*, June 11, 1917, page 1.
20. Ibid.
21. Ibid.; *Butte Daily Post*, June 11, 1917, page 8.
22. Ibid.

23. Author interview with Al Hooper, September 4, 2003.

24. The telegraph from J. D. Moore's mother (Mrs. M. D. Murphy) is on file at the Montana Historical Society Archives, Helena.

25. By Clarence Marthey's account, Moore, a few hours before rescue, "got a candle going and read to us what he had written." Marthey admits that "we lost all track of time," and it appears unlikely that Moore would have been able to light a candle in the hours immediately preceding the rescue. The air quality had been too poor to sustain a candle twenty-four hours earlier and would have continued to deteriorate. Marthey's account also showed confusion about other events. The two most likely scenarios are that (1) Moore actually read his will to the men *before* they lost light or that (2) Moore recited the contents of the will to his men from memory. See Marthey account, *Anaconda Standard,* June 12, 1917, page 11.

26. Ibid.

Chapter 17. *"We're Dying in Here"*

1. *Anaconda Standard,* June 12, 1917, page 11.

2. The dialogue is drawn from articles in the *Butte Daily Post,* June 11, 1917, page 1, and the *Anaconda Standard,* June 12, 1917, page 11.

3. *Anaconda Standard,* June 12, 1917, page 11; *Butte Daily Post,* June 11, 1917, page 7.

4. See Daniel Harrington, Bureau of Mines/Department of Interior, *Lessons from the Granite Mountain Shaft Fire, Butte (Bureau of Mines Accident Report),* 1922, page 30.

5. *Butte Miner,* June 13, 1917, page 14.

6. *Anaconda Standard,* June 12, 1917, page 1.

7. *Butte Daily Post,* June 11, 1917, pages 1, 7.

8. *Anaconda Standard,* June 12, 1917, page 8.

9. *Butte Daily Post,* June 11, 1917, page 7.

10. Ibid.

11. *Anaconda Standard,* June 12, 1917, pages 1, 11.

12. *Butte Daily Post,* June 11, 1917, page 1; *Bureau of Mines Accident Report,* page 24.

13. *Butte Daily Post,* June 11, 1917, 11.

14. *Bureau of Mines Accident Report,* page 24. *Anaconda Standard,* June 12, 1917, page 11. The *Anaconda Standard* account gives the impression that the doctor did not have a breathing apparatus to begin with. This seems unlikely,

given the steps taken by other rescuers and the efforts to secure apparatus for the trapped men.

15. The *Bureau of Mines Accident Report* states that there were eight men in the Moore party—two of whom died—without ever explaining the discrepancies with the contemporaneous accounts. Contemporaneous accounts refer consistently to ten men in the party—four of whom were killed. However, two of the men in these accounts are referred to as "unidentified," making it possible that the numbers were mistaken in the general confusion on the surface. See *Bureau of Mines Accident Report,* page 28.

16. *Anaconda Standard,* June 12, 1917, page 11.

17. Ibid.

18. *Butte Daily Post,* June 11, 1917, page 7.

19. *Anaconda Standard,* June 12, 1917, page 11; *Butte Daily Post,* June 11, 1917, page 7; *Anaconda Standard,* June 13, 1917, page 10; *Bureau of Mines Accident Report,* page 30.

20. *Anaconda Standard,* June 12, 1917, page 11.

21. Ibid., page 1.

22. Ibid.

23. *Butte Daily Post,* June 11, 1917, page 1.

24. *Anaconda Standard,* June 12, 1917, page 4.

25. *Butte Miner,* June 14, 1917, page 14.

26. Ibid., page 8; *Butte Daily Post,* June 11, 1917, page 1.

27. *Anaconda Standard,* June 12, 1917, page 11.

28. The original copy of the Western Union Telegram from J.D. Moore's mother—Mrs. M. D. Murphy—is on file at the Montana Historical Society Archives, Helena.

29. The original copy of the letter dictating the North Butte Mining Company's return telegraph to J.D. Moore's mother is on file at the Montana Historical Society Archives, Helena.

30. *Butte Daily Post,* June 11, 1917, page 7.

Chapter 18. "Point of Eruption"

1. *Butte Daily Post,* June 12, 1917, pages 1, 3; Dunne cited by Arnon Gutfeld, "The Speculator Disaster in 1917: Labor Resurgence at Butte, Montana," *Arizona and the West,* vol. 11, no. 1 (Spring 1969), page 31.

2. *Third Biennial Report of the Department of Labor,* 1917–1918, pages 17–18.

3. Handbill entitled "Miners, Attention! Let's Have a Union!," June 11, 1917, viewed on microfilm at the Montana Historical Society Library, Helena.

4. The demands of the Metal Mine Workers' Union are quoted from a document known as the *Glasser File,* pages 8–26. The *Glasser File,* in turn, cites Department of Labor File no. 33/493. The *Glasser File* was a report prepared between 1933 and 1934 by Ira Glasser, a civil servant at the U.S. Department of Justice. Glasser's report, which was not published, was commissioned by President Franklin Roosevelt. Roosevelt feared that the severe economic conditions of the Depression might lead to domestic unrest and wanted a study of the federal government's past handling of such situations. One focus of Glasser's study was Butte during and after World War I. A copy of the *Glasser File* is available at the Butte-Silver Bow Public Archives.

5. *Glasser File,* page 18, quoting strikers.

6. See *Metal Mine Workers Strike Bulletin,* undated, viewed on microfilm at the Montana Historical Society Library, Helena. While Manus Duggan's daughter, Manus Dugan Banko, does not discount this claim out of hand, she does not know if it is true. Author interview, November 9, 2004.

7. Ibid.

8. *Anaconda Standard,* June 11, 1917, page 9.

9. See Burton K. Wheeler, *Yankee from the West* (1962), page 137.

10. Daniel Harrington, Bureau of Mines/Department of Interior, *Lessons from the Granite Mountain Shaft Fire, Butte (Bureau of Mines Accident Report),* 1922, page 46. The report did allow that some men might have been killed in other parts of the mine after bulkheads turned them back: "[T]he existence of bulkheads on the 2,400 and 2,800 levels to High Ore connection may have caused some men to retreat and later on to lose their lives."

    There are at least three documents showing the North Butte Mining Company's efforts to change the text of the original Bureau of Mines report. On March 30, 1918, Norman Braley, the North Butte's general manager, sent a letter to Robert Linton, the North Butte's vice president, "forwarding herewith Mr. Harrington's report on the Granite Mountain fire." The letter includes a long list of objections to the report. On April 9, 1918, Linton writes back to Braley and directs him to "discuss the report with Mr. Harrington when you see him and when I come to Butte again in June we can go into the matter more thoroughly and probably arrange a modification of the report which would be acceptable." Finally, on May 14, 1918, Linton writes Braley to report on a meeting in Washington with "Mr. Rice," apparently Harrington's boss. "It might save some time," writes Linton, "if you and Mr. Harrington would have some discussion of the points we object to and then when

I come out we can give the matter a final going over." The above letters are on file at the Montana Historical Society Library, Helena.

11. George Harrison, *The I.W.W. Trial, Story of the Greatest Trial in Labor's History by One of the Defendants* (1969, original edition, 1919), pages 118–119.

12. Ibid., page 81.

13. See, e.g., *Butte Daily Post,* June 9, 1917, page 3; *Anaconda Standard,* June 11, 1917, page 9; *Butte Miner,* June 10, 1917, page 10; *Butte Miner,* June 10, 1917, page 12.

14. An 1897 state law required all mines of greater than 100 feet to maintain escapement shafts. It specifically envisioned the use of neighboring mines as escape routes, providing that "the right to use the outlet through such contiguous mine in all cases when necessary, or in cases of accident must be secured and kept in force." The law further required that "the exit, escapement shaft, raise, or opening provided for . . . must be of sufficient size as to afford an easy passage way . . ." The statute also required signage to mark escape routes. *Revised Codes of Montana, 1907,* Section 8541.

15. *Bureau of Mines Accident Report,* page 32.

16. *Metal Mine Workers Strike Bulletin,* undated, viewed on microfilm at the Montana Historical Society Library, Helena.

17. *Anaconda Standard,* June 21, 1917, page 6.

18. *Third Biennial Report,* page 5.

19. Ibid., page 17.

20. Letter from U.S. Attorney Burton K. Wheeler to the attorney general of the United States, August 21, 1917, Department of Justice File No. 186701-27, cited in the *Glasser File,* page 45.

21. *Glasser File,* pages 9–10. Glasser compares figures from the Bureau of Labor Statistics on wholesale copper prices against a Department of Labor table of wage rates paid in Butte mines for the corresponding period.

22. Jerry W. Calvert, *The Gibraltar: Socialism and Labor in Butte, Montana, 1895-1920,* page 106; Gutfeld, "The Speculator Disaster in 1917: Labor Resurgence at Butte, Montana," page 35.

23. The IWW is still alive today, at least as a Web site. The IWW Constitution is quoted from iww.org. Recent IWW activities, according to the Web site, include the formation of an affiliate union at one Starbucks store (at Thirty-sixth Street and Madison Avenue in New York City). See also Gutfeld, "The Speculator Disaster in 1917: Labor Resurgence at Butte, Montana," page 30.

24. The mayor was former Unitarian minister Luis Duncan, who would ultimately be impeached in the controversial crackdown following the 1914 destruction of the Miners' Union Hall. See Calvert, pages 52–53. Butte was not

the only place to witness conflict between socialists and the IWW. In his 1913 book *American Syndicalism,* John Graham Brooks notes (page 73) that a "prolific IWW literature has more acrid abuse of the many prominent socialist leaders than anything appearing in capitalistic sheets."

25. *Glasser File,* page 20.

26. Paul F. Brissenden, "The Butte Miners and the Rustling Card," *The American Economic Review,* vol. 10, no. 4 (December 1920), page 758.

27. Michael P. Malone, Richard B. Roeder, and William L. Lang, *Montana: A History of Two Centuries* ("Malone et al., *A History of Two Centuries*"), revised edition, 1991, page 271.

28. See *Anaconda Standard,* June 17, 1917, page 1; Calvert, page 105.

29. *Miners' and Electrical Workers' Joint Strike Bulletin,* undated, viewed on microfilm at the Montana State Historical Society Library, Helena.

30. *Anaconda Standard,* June 13, 1917, page 1.

31. Ralph Chaplin, *Wobbly—The Rough-and-Tumble Story of an American Radical* (1948), page 208.

32. Wheeler, page 137; see also Arnon Gutfeld, *The Butte Labor Strikes and Company Retaliation during World War I,* master's thesis, University of Montana (1967), page 13.

33. *Butte Daily Post,* June 12, 1917, page 1; *Anaconda Standard,* June 16, 1917, page 1.

34. *Anaconda Standard,* June 21, 1917, page 6.

35. *Butte Miner,* June 15, 1917, page 12.

36. *Anaconda Standard,* June 17, 1917, page 1.

37. *Anaconda Standard,* June 13, 1917, page 1.

38. *North Butte Mining Company, Annual Report for the Year Ended December 31, 1916,* on file at the Montana Historical Society Archives, Helena; Michael P. Malone, *The Battle for Butte: Mining and Politics on the Northern Frontier, 1864–1906 (1981),* pages 166–167, 201.

Chapter 19. "For You and the Child"

1. *Butte Daily Post,* June 11, 1917, page 8.

2. *Anaconda Standard,* June 14, 1917, page 7.

3. According to the *Anaconda Standard,* June 12, 1917, page 11, a miner named Mike Sullivan (with the J.D. Moore group) "refused to drink the coffee, thinking an attempt was being made to poison him."

4. Cobb account, *Anaconda Standard,* June 13, 1917, page 12.

5. Letter from Norman B. Braley to the Carnegie Hero Fund Commission, July

25, 1917, on file in the collections of the Montana Historical Society Archives, Helena.

6. *Anaconda Standard,* June 11, 1917, page 1.

7. *Butte Miner,* June 11, 1917, page 1.

8. *Anaconda Standard,* June 13, 1917, page 12.

9. See Josiah James's account in an article by Andrea McCormick, "Men Struggle for Life in Gas-Filled Mines," *Montana Standard,* June 1, 1980, page 23; see also Wilfred LaMontague: "He [Duggan] got out first and must have tried to make the Rainbow." Cited in the *Butte Miner,* June 11, 1917, page 8.

10. *Butte Miner,* June 11, 1917, page 1.

11. *Anaconda Standard,* June 14, 1917, page 1.

12. See *Anaconda Standard,* June 14, 1917, pages 1, 7. An earlier, false account of finding Duggan's body had been published in the *Butte Daily Post* on Wednesday, June 13, 1917, page 7. According to this account, the body was found on the 2,600 level "in a sitting posture, just as if the young nipper, after giving up the fight for life as vain, had calmly sat down to die."

13. *Butte Miner,* June 11, 1917.

14. Author interviews with Manus Dugan Banko (Manus Duggan's daughter), March 11, 2004, and June 10, 2004.

15. *Butte Daily Post,* June 11, 1917, page 1.

16. *Butte Miner,* June 12, 1917, page 9.

17. *Anaconda Standard,* June 12, 1917, page 1.

18. *Anaconda Standard,* June 14, 1917, page 1.

19. *Anaconda Standard,* June 12, 1917, page 1.

20. Accounts of the helmet men finding Duggan are drawn from the *Butte Daily Post,* June 14, 1917, page 1, and the *Anaconda Standard,* June 15, 1917, page 7.

Chapter 20. *"Dupes and Catspaws"*

1. *Anaconda Standard,* June 15, 1917, page 1; Daniel Harrington, Bureau of Mines/Department of Interior, *Lessons from the Granite Mountain Shaft Fire, Butte (Bureau of Mines Accident Report),* 1922, page 21. The precise number of dead remains unclear. Butte historian James D. Harrington completed an analysis of all available documentation and concluded that "at least 167" men died. See "A Reexamination of the Granite Mountain-Speculator Fire," *Montana* (Autumn 1998), pages 62–69.

2. Technically, the electricians were on strike against Montana Power—not Anaconda. In reality, though, this is a distinction without a difference. Though

legally separate and independent companies, Montana Power and Anaconda shared management and other resources. Both were referred to as "The Company," and the pair was sometimes called the "Siamese Twins." See generally Jerry W. Calvert, *The Gibraltar: Socialism and Labor in Butte, Montana, 1895-1920* (1988), pages 105-106.

3. *Butte Daily Post,* June 19, 1917, page 1.

4. *Butte Miner,* June 15, 1917, page 1.

5. *Anaconda Standard,* June 17, 1917, page 1.

6. *Anaconda Standard,* June 22, 1917, page 1.

7. *Miners' and Electrical Workers' Joint Strike Bulletin,* undated, viewed on microfilm at the Montana Historical Society Library, Helena.

8. Several sources discuss the link between Wheeler and the formation of a labor newspaper in Butte. The most detailed is a master's thesis by Guy Ole Halverson, *The Butte Bulletin: A Newspaper of Montana Progressivism* (1967), pages 14-17. In researching his thesis, Halverson apparently corresponded directly with Wheeler on this topic, citing among his sources two personal letters from Wheeler.

9. Wheeler letter to the U.S. attorney general, August 21, 1917, cited in the *Glasser File,* pages 43-46. The *Glasser File* was a report prepared between 1933 and 1934 by Ira Glasser, a civil servant at the U.S. Department of Justice. Glasser's report, which was not published, was commissioned by President Franklin Roosevelt. Roosevelt feared that the severe economic conditions of the Depression might lead to domestic unrest, and wanted a study of the federal government's past handling of such situations. One focus of Glasser's study was Butte during and after World War I. A copy of the *Glasser File* is on file at the Butte-Silver Bow Public Archives.

10. According to Halverson, one of the other attorneys who helped to launch the *Butte Bulletin* was Butte Federal District Judge George Bourquin—whose controversial 1918 decision in the Ves Hall case would deepen the political crisis in Montana (discussed in Chapter 22). See Halverson, page 14. *Butte Bulletin* editor William Dunne would become increasingly radical throughout his career. In 1918, he would run successfully (technically, as a Democrat) for the Montana House of Representatives. His writings, however, make it clear that his leanings were socialistic. Guy Ole Halverson, *The Butte Bulletin: A Newspaper of Montana Progressivism* (thesis, 1967), pages 44-45, on file at the University of Montana's Mansfield Library in Missoula.

11. *Anaconda Standard,* June 14, 1917, page 7.

12. *Anaconda Standard,* June 16, 1917, page 7; see also *Butte Miner,* June 16, 1917, page 9.

13. Andrea McCormick, "Woman Inherits Mine's Heroic Legacy," *Montana Standard,* November 5, 1978, page 22.

14. *Anaconda Standard,* June 16, 1917, page 7; see also *Butte Miner,* June 16, 1917, page 9.

15. John 15:13.

16. *Anaconda Standard,* June 18, 1917, page 1.

17. *Third Biennial Report of the Department of Labor, 1917–1918,* page 19.

18. *Glasser File,* pages 1, 30.

19. See *Anaconda Standard,* June 21, 1917, page 1, and June 23, 1917, page 1; *Glasser File,* page 64; Calvert, page 107.

20. See *Butte Daily Post,* June 16, 1917, page 2.

21. *Miners' and Electrical Workers' Joint Strike Bulletin,* undated, viewed on microfilm at the Montana Historical Society Library, Helena. Michael P. Malone, Richard B. Roeder, and William L. Lang, *Montana: A History of Two Centuries,* revised edition, 1991, page 274.

22. *Third Biennial Report,* page 19; Calvert, page 107.

23. *Butte Miner,* June 14, 1917, page 1.

24. *Anaconda Standard,* June 16, 1917, page 1.

25. *Miners' and Electrical Workers' Joint Strike Bulletin,* undated, viewed on microfilm at the Montana Historical Society Library, Helena.

26. *Anaconda Standard,* June 15, 1917, page 7.

27. *Miners' and Electrical Workers' Joint Strike Bulletin,* July 12, 1917, viewed on microfilm at the Montana Historical Society Library, Helena.

28. See, e.g., Malone, Roeder, and Lang, page 274.

29. *Miners' and Electrical Workers' Joint Strike Bulletin,* undated, viewed on microfilm at the Montana Historical Society Library, Helena.

30. Transcript of interview with Burton K. Wheeler by John Paxon in Washington, D.C., on February 7, 1972, page 6. Transcript on file at the Montana Historical Society Library, Helena.

31. *Anaconda Standard,* June 21, 1917, page 6.

32. *Butte Daily Post,* June 14, 1917, page 1.

33. *Anaconda Standard,* June 14, 1917, page 7.

34. *Butte Miner,* June 16, 1917, page 1; *Anaconda Standard,* June 14, 1917, page 1; *Butte Daily Post,* June 13, 1917.

35. *Miners' and Electrical Workers' Joint Strike Bulletin,* July 4, 1917.

36. Letter from Governor W. B. Stewert to Colonel W. E. Elliss, December 8, 1917, printed in full in the *Glasser File,* pages 73–75.

37. *Glasser File,* page 64.

38. *Third Biennial Report,* page 17.

39. *Anaconda Standard,* June 16, 1917, page 1.

40. *Anaconda Standard,* June 17, 1917, page 1; *Anaconda Standard,* June 18, 1917, page 1.

41. Wheeler letter to the attorney general of the United States, August 21, 1917, Department of Justice File No. 186701-27, cited by the *Glasser File,* pages 43–46.

42. Arnon Gutfeld, "The Speculator Disaster," *Arizona and the West: A Quarterly Journal of History* (Spring 1969), page 32; Calvert, page 79.

43. Wheeler, page 70.

44. Calvert, page 75.

45. Gutfeld, "The Speculator Disaster," page 35; *Anaconda Standard,* June 13, 1917, page 1; *Anaconda Standard,* June 18, 1917, page 1.

46. Calvert, page 106.

47. *Miners' and Electrical Workers' Joint Strike Bulletin,* July 12, 1917, viewed on microfilm at the Montana Historical Society Library, Helena.

48. Ibid.

49. *Miners' and Electrical Workers' Joint Strike Bulletin,* July 16, 1917, viewed on microfilm at the Montana Historical Society Library, Helena; Calvert, page 106.

50. Gutfeld, "The Speculator Disaster," page 37.

51. *Glasser File,* page 29, footnote 2.

52. George R. Tompkins, *The Truth about Butte* (1917), page 57.

53. *Glasser File,* page 32.

54. The description of the IWW meeting and dialogue is found in Ralph Chaplin's *Wobbly—The Rough-and-Tumble Story of an American Radical* (1948), pages 208–209.

55. Andrea McCormick, "Woman Inherits Mine's Heroic Legacy," *Montana Standard,* November 5, 1978, page 22.

Chapter 21. "Others Take Notice"

1. Ralph Chaplin, *Wobbly—The Rough-and-Tumble Story of an American Radical* (1948), pages 195–196, 208.

2. Dunne cited by Arnon Gutfeld, "The Murder of Frank Little: Radical Labor Agitation in Butte, Montana, 1917," *Labor History* (Spring 1969), page 189.

3. Michael P. Malone, Richard B. Roeder, and William L. Lang, *Montana: A History of Two Centuries,* revised edition, 1991, page 274.

4. *Anaconda Standard,* August 2, 1917, page 2; see also Gutfeld, "The Murder

of Frank Little," pages 183–184; Burton K. Wheeler, *Yankee from the West* (1962), page 139.

5. Little cited by Gutfeld, "The Murder of Frank Little," page 185.

6. *Anaconda Standard,* August 2, 1917, page 2.

7. *Butte Miner,* cited by Gutfeld, "The Murder of Frank Little," page 183.

8. *Butte Miner,* July 21, 1917, cited by Gutfeld, "The Murder of Frank Little," page 184. Regarding Anaconda's contacts with Wheeler, see Wheeler's testimony before the Montana Council of Defense, May 31, June 1–2, 1918, transcript page 363, on file in the Montana Historical Society Archives, Helena. See also Wheeler, pages 139–140; Wheeler statement in the *Anaconda Standard,* August 2, 1917, page 3.

9. Arnon Gutfeld, *Montana's Agony* (1979), page 26.

10. *Miners' and Electrical Workers' Joint Strike Bulletin,* July 16, 1917, viewed on microfilm at the Montana Historical Society Library, Helena.

11. Isaac F. Marcosson, *Anaconda* (1957), pages 102–103.

12. Wheeler, page 140.

13. Interview of Burton K. Wheeler by John Paxon, February 7, 1972, page 6. Transcript on file at the Montana Historical Society Archives, Helena.

14. Espionage Act, cited by Wheeler, page 151.

15. Wheeler, page 152.

16. Ibid., pages 139–140.

17. The description of events and the dialogue are reported in the *Anaconda Standard,* August 2, 1917, page 1.

18. Waldemar Kaiyala, "My Memories of Butte Montana," handwritten document on file at the Butte-Silver Bow Public Archives.

19. Gutfeld, "The Murder of Frank Little," page 178.

20. Frank Little's death certificate is on file at the Butte-Silver Bow Public Archives.

21. *Strike Bulletin,* August 2, 1917, viewed on microfilm at the Montana Historical Society Library, Helena.

22. *Anaconda Standard,* August 2, 1917, page 3.

23. *Helena Independent,* August 2, 1917, page 4.

24. *Anaconda Standard,* August 2, 1917, pages 1, 6.

25. *Butte Miner,* August 2, 1917, page 4.

26. Even today, Montana's Highway Patrol Officers wear the numbers 3-7-77 on their sleeve patches, a harkening back to an earlier system of law and order.

27. *Anaconda Standard,* August 2, 1917, page 1.

28. Arnon Gutfeld, *The Butte Labor Strikes and Company Retaliation During World War I,* master's thesis (1967), page 33, on file at the University of Montana's Mike and Maureen Mansfield Library in Missoula.

29. *Strike Bulletin,* August 2, 1917, viewed on microfilm at the Montana Histori-
     cal Society Library, Helena.

30. For speculation on Little's murderers, see Wheeler, pages 141–142; Arnon
     Gutfeld, "The Murder of Frank Little," pages 34–38; *Anaconda Standard,*
     August 2, 1917, page 7.

31. *Strike Bulletin,* August 2, 1917.

32. Joseph Kinsey Howard, *Montana: High, Wide, and Handsome* (1943), page 89.

33. Mike Byrnes and Les Rickey, *The Truth about the Lynching of Frank Little*
     (2003), pages 106–109.

34. Interview of Burton K. Wheeler by John Paxon, February 7, 1972, page 7.
     Transcript on file at the Montana Historical Society Archives, Helena.

Chapter 22. "Spy Fever"

1.   Arnon Gutfeld, "The Murder of Frank Little: Radical Labor Agitation in
     Butte, Montana, 1917," *Labor History* (Spring 1969), page 189; Arnon Gut-
     feld, *Montana's Agony* (1979), page 50; *Glasser File,* pages 49–50. The
     Glasser File was a report prepared between 1933 and 1934 by Ira Glasser, a
     civil servant at the U.S. Department of Justice. Glasser's report, which was
     not published, was commissioned by President Franklin Roosevelt. Roosevelt
     feared that the severe economic conditions of the Depression might lead to
     domestic unrest, and wanted a study of the federal government's past han-
     dling of such situations. One focus of Glasser's study was Butte during and
     after World War I. A copy of the *Glasser File* is on file at the Butte-Silver Bow
     Public Archives.

2.   Waldemar Kaiyala, "My Memories of Butte Montana," handwritten docu-
     ment on file at the Butte-Silver Bow Public Archives.

3.   Arnon Gutfeld, "The Murder of Frank Little: Radical Labor Agitation in
     Butte, Montana, 1917," *Labor History* (Spring 1969), pages 177, 189.

4.   See, e.g., the *Anaconda Standard,* July 25, 1917, pages 1, 10.

5.   *Glasser File,* pages 33–36.

6.   *Strike Bulletin,* July 23, 1919. Viewed on microfilm at the Montana Historical
     Society Library, Helena.

7.   Ibid.

8.   *Strike Bulletin,* July 31, 1917. Viewed on microfilm at the Montana Historical
     Society Library, Helena.

9.   *Glasser File,* pages 68–70; Arnon Gutfeld, "The Murder of Frank Little,"
     page 188.

10. *Glasser File,* page 51.

11. *Anaconda Standard,* August 11, 1915, page 1.

12. *Helena Independent,* August 11, 1917, page 1. See also the *Glasser File,* pages 61–62.

13. *Glasser File,* page 65.

14. Clayton D. Laurie and Ronald H. Cole, *The Role of Federal Military Forces in Domestic Disorders, 1877–1945* (1997), page 223. For a map showing states where interventions occurred, see pages 226–227.

15. *Glasser File,* page 73. *Anaconda Standard,* December 19, 1917, page 1.

16. Burton K. Wheeler, *Yankee from the West* (1962), page 142.

17. Wheeler testimony before the Montana Council of Defense, June 4–5, 1918, page 1122, on file at the Montana Historical Society Archives, Helena; Gutfield, *Montana's Agony,* page 60.

18. Wheeler, page 144.

19. For "despised language" see *Helena Independent,* August 31, 1917, page 4. For the Apache story see *Helena Independent,* August 8, 1917, page 1. For a comprehensive survey and analysis of Campbell's writing during this period, see Charles S. Johnson, *An Editor and a War: Will A. Campbell and the Helena Independent* (1977), master's thesis on file at the University of Montana's Mansfield Library, Missoula, pages 55–128.

20. *Helena Independent,* September 1, 1917, page 1.

21. *Helena Independent,* October 18, 1917, page 1.

22. *Helena Independent,* November 2, 1917, pages 1–2.

23. *Helena Independent,* October 18, 1917, page 1.

24. As for airplanes specifically, B.K. Wheeler noted, "It must be remembered that the airplane was an excitingly new and mysterious machine in the West. While Americans living on the coastal areas feared submarine attack, inland Westerners had no trouble at all worrying about invasion from the air." See Wheeler, pages 143–144.

25. The most comprehensive examination of the Montana Council of Defense is a 1966 master's thesis by Nancy Rice Fritz entitled *The Montana Council of Defense,* on file at the University of Montana's Mansfield Library in Missoula. For background on President Wilson's formation of the state councils, see Fritz, pages 6–10.

26. Ibid., pages 11–23.

27. Ibid., page 11.

28. Fritz, pages 18, 23.

29. K. Ross Toole, *20th Century Montana: A State of Extremes* (1972), page 176. Michael P. Malone, Richard B. Roeder, and William L. Lang, *Montana: A History of Two Centuries,* revised edition, 1991, pages 268–269.

30. Fritz, pages 16–18, 23.

31. For "get that paunch down" see *Helena Independent,* May 11, 1917, page 11. For "farm during your vacation" see *Helena Independent,* May 4, 1917, page 4. For "booze and efficiency" see *Helena Independent,* February 20, 1917, page 4. See also Johnson, Chapters 4, 5.

32. For "an excellent thing" see *Helena Independent,* July 13, 1917, page 4. For "intern camps" see *Helena Independent,* May 16, 1917, page 4.

33. *Helena Independent,* August 2, 1917, page 4. See also Johnson, pages 82–86, 88–89.

34. Johnson, pages 88–89.

35. *Helena Independent,* August 30, 1917, page 4; Johnson, pages 102–103.

36. *Helena Independent,* August 6, 1917, page 4; Johnson, pages 102–103.

37. Fritz, pages 24–25.

38. K. Ross Toole, *20th Century Montana: A State of Extremes* (1972), page 169; Wheeler, pages 153–154; Johnson, page 119.

39. According to Wheeler's autobiography, Judge Bourquin had taken the questionable step of actually encouraging Wheeler to "send some of those sedition cases up to me and I'll take care of them. I'm in a stronger position than you are." Wheeler, page 152.

40. Toole, page 168; Johnson, page 119.

41. Wheeler, page 168; Johnson, page 120.

42. *Helena Independent,* January 28, 1918. Johnson, page 121.

43. Fritz, page 25, citing the governor's opening proclamation to the Extraordinary Session.

Chapter 23. *"What Will Become of Them"*

1. *Butte Miner,* June 10, 1917, page 14.

2. Author interview with Manus Dugan Banko, November 9, 2004. See also Letter from Norman Braley to Madge B. Duggan, June 11, 1919, requesting that the bill for the stone be sent to the North Butte Mining Company, on file at the Montana Historical Society Archives, Helena.

3. *Butte Daily Post,* June 12, 1917, page 2.

4. Letter from the Servian Bocalian Society of Butte to the North Butte Mining Company, September 13, 1917, on file at the Montana Historical Society Archives, Helena.

5. Benefits described by Montana State Auditor Keating in the *Anaconda Standard,* June 10, 1917, page 11.

6. Jerry W. Calvert, *The Gibraltar* (1988), page 139. See also Michael P. Ma-

lone, Richard B. Roeder, and William L. Lang, *Montana: A History of Two Centuries,* revised edition, 1991, page 260.

7. Letters from Mrs. B. Thompson to the North Butte Mining Company, July 12, 1917, and August 10, 1917, on file at the Montana Historical Society Archives, Helena.

8. *Metal Mine Workers Strike Bulletin,* undated, viewed on microfilm at the Montana Historical Society Library, Helena.

9. See, for example, *Anaconda Standard,* June 10, 1917, page 1. The figures on the workman's compensation were compiled in a document by Jackie Corr, August 1, 2001, on file at the Montana Historical Society Archives, Helena. Weak as it was, the 1915 Montana Workman's Compensation law at least ensured some payments to some families. Without the law, it is unlikely that any miners would have received anything. According to former Governor A. E. Spriggs, a member of the state mining board, "If it were not for the law the families of those killed in this latest disaster would not be able to collect a cent, as we feel that the North Butte Mining Company is absolutely blameless in this matter." *Anaconda Standard,* June 11, 1917, page 9.

10. Jerry W. Calvert, *The Gibraltar* (1988), page 139.

11. Unless otherwise indicated, information about Madge Duggan is drawn from three author interviews with her daughter, Manus Dugan Banko, on March 11, 2004; June 10, 2004; and November 9, 2004.

12. *Butte Miner,* June 10, 1917, page 12; *Butte Daily Post,* June 13, 1917, page 9.

13. *Anaconda Standard,* June 10, 1917, page 11.

14. Information about Con O'Neill and his family comes from May 17, 2004, interviews with his granddaughters, Sally Wolahan Rasmussen and Jeanne Weber.

15. Author interview with Jeanne Weber, May 17, 2004.

16. Ibid.

17. Mary Murphy, *Mining Cultures: Men, Women, and Leisure in Butte, 1914–41,* pages 19–20. Murphy's book is an excellent resource on the lives of women in Butte. The quotation is from Ann Pentilla, cited by Murphy, page 19.

18. Election results are listed in the *Anaconda Standard,* November 7, 1918, page 6.

19. *Anaconda Standard,* November 6, 1918, page 1.

20. Piece C. Mullen and Michael L. Nelson, "Montanans and the Influenza Epidemic," *Montana: the Magazine of Western History* (Spring 1987), pages 50–61. See also "Saloon Ban Lifted for Premature Celebration," *Anaconda Standard,* November 8, 1918, page 10.

21. See, e.g., *Butte Daily Post,* November 6, 1918, page 1.

22. Madge beat her opponent, G. W. Deniger, by 5,873 to 5,330 votes, or 52 to 48 percent. *Anaconda Standard,* November 3, 1920, page 5.

Chapter 24. *"Some Little Body of Men"*

1. Wheeler testimony cited by Arnon Gutfeld, *Montana's Agony* (1979), page 59. Wheeler Testimony Before the Montana Council of Defense, May 31–June 2, 1918, pages 1289–1290.

2. Stewart speech printed in full in the *Helena Independent,* February 15, 1918, page 2. See also K. Ross Toole, *20th Century Montana: A State of Extremes* (1972), page 173; Nancy Rice Fritz, *The Montana Council of Defense* (1966), master's thesis on file at the University of Montana's Mansfield Library, Missoula, page 24.

3. Act of the Extraordinary Session of the Fifteenth Legislative Assembly Creating the Montana Council of Defense, 1918, Sections 2, 4, 5, 6, and 10. Published in *Laws Passed by the Extraordinary Session of the Fifteenth Legislative Assembly* (Helena, State Publishing Co., no date), pages 14–16.

4. *Helena Independent,* February 3, 1918, page 4; Charles S. Johnson, *An Editor and a War: Will A. Campbell and the Helena Independent* (1977), master's thesis on file at the University of Montana's Mansfield Library, Missoula, page 122.

5. Codified in *Revised Codes of Montana, 1921,* Section 10737 et seq.

6. *Congressional Record,* 65th Congress, 2nd Session, 1918, LVI, Part 5, 4695; Robert Emlyn Evans, Montana's Role in the Enactment of Legislation Designed to Suppress the Industrial Workers of the World (1963), master's thesis on file at the University of Montana's Mansfield Library, Missoula, page 91.

7. *Anaconda Standard,* May 28, 1918, page 6; Gutfeld, *Montana's Agony,* page 47.

8. The Criminal Syndicalism Act was codified as Section 10740 et seq., *Revised Codes of Montana, 1921.* See also Johnson, pages 123–124; Michael P. Malone, Richard B. Roeder, and William L. Lang, *Montana: A History of Two Centuries,* revised edition, 1991, page 276.

9. Johnson, pages 124–126; Burton K. Wheeler, *Yankee from the West* (1962), page 155.

10. *Helena Independent,* February 27, 1918, page 4; Johnson, pages 124–125.

11. Nancy Rice Fritz, *The Montana Council of Defense* (1966), master's thesis on file at the University of Montana's Mansfield Library in Missoula, page 35; Johnson, page 131.

12. Montana Council of Defense, Order Number One, March 15, 1918 (appended in Fritz, page 129); Johnson, pages 131–132.

13. Montana Council of Defense, Order Number Twelve, August 12, 1918 (appended in Fritz, page 137).

14. Kurt Wetzel, *The Making of an American Radical: William Dunne in Butte* (1970), master's thesis on file at the University of Montana's Mansfield Library, Missoula, page 26. Johnson, pages 175–178. Burton K. Wheeler claimed to have been an early supporter and even an investor in the *Butte Bulletin.* See Wetzel, page 26; Arnon Gutfeld, *Montana's Agony* (1979), pages 83–86.

15. Fritz, pages 91–95.

16. Montana Council of Defense, Order Number Three, April 22, 1918 (appended in Fritz, page 131); Fritz, page 94.

17. *Helena Independent,* August 16, 1917, page 4; Johnson, page 98.

18. Fritz, pages 96–97. See also Toole, *20th Century Montana,* page 188.

19. Cited by Fritz, page 101.

20. Montana Council of Defense, Order Number Seven, May 28, 1918 (appended in Fritz, page 133). The rules of procedure are found under Order Number Eight, May 28, 1917 (appended in Fritz, page 134).

21. See, e.g., Toole, *20th Century Montana,* Chapter 7: "The Inquisition;" Gutfeld, *Montana's Agony,* Chapter 5: "Panic and Inquisition."

22. Jerry W. Calvert, *The Gibraltar* (1988), page 113.

23. See Richard L. Neuberger, "Wheeler of Montana," *Harper's Magazine* (May 1940), page 615; *Anaconda Standard,* October 10, 1918, page 1.

24. Johnson, page 144.

25. Wheeler, page 159. See also *Helena Independent,* May 30, 1918, pages 1, 7.

26. *Helena Independent,* June 6, 1918, page 1.

27. Wheeler, cited by Fritz, page 68; see also Toole, page 183.

28. Wheeler Testimony Before the Montana Council of Defense ("Wheeler Testimony"), May 31–June 2, 1918, pages 352–354. Testimony on file at the Montana Historical Society Archives, Helena.

29. Fritz, page 70.

30. Wheeler Testimony, pages 368–373; Toole, page 183.

31. Anaconda men testifying included General Manager John Gillie and Roy S. Alley. *Helena Independent,* June 1, 1918, page 4.

32. On Anaconda's close connections to the Montana Council of Defense, see Fritz, page 53; on Anaconda's testimony against Wheeler, see Wheeler, page 159.

33. Toole, pages 178–179. Evans cited by Toole, page 183. *Helena Independent,* June 5, 1918, page 1.

34. Wheeler cited by Gutfeld, page 59. Wheeler Testimony, pages 1289–1291.

35. The *Helena Independent,* June 6, 1918, page 7, printed the council resolution in its entirety.

36. Gutfeld, *Montana's Agony,* page 67; Fritz, page 85.

37. Fritz, page 78. The *Seattle Post-Intelligencer* ad (from October 5, 1918) is cited by H. C. Peterson and Gilbert C. Fite, *Opponents of War, 1917–1918* (1957), page 142.

38. Interrogation cited by Fritz, pages 79–81.

39. Fritz, page 81; Malone, Roeder, and Lang, page 278.

40. Gutfeld, *Montana's Agony,* pages 99–100.

41. The full text of the Ford letter is printed in Appendix V of Fritz's thesis.

42. Geoffrey R. Stone, *Perilous Times: Free Speech in Wartime from the Sedition Act of 1798 to the War on Terrorism* (2004), page 535.

43. Evans, pages 123–124; Gutfeld, *Montana's Agony,* page 46.

44. Fritz, page 108.

45. Stone, pages 12, 230, 530.

46. Ibid., page 230.

47. Wheeler, page 158. Except where noted otherwise, events and quotations in this section are drawn from Wheeler's autobiography, *Yankee from the West* (1962), pages 160–164.

48. *Helena Independent,* October 22, 1917, page 4.

49. *Anaconda Standard,* October 3, 1918, page 4.

50. In a meeting with Thomas Gregory, the United States attorney general, Wheeler was offered a federal judgeship in Panama. "If you're going to deport me, you'd better make it Siberia," replied Wheeler bitterly. "I understand people don't live very long down in Panama." Wheeler also turned down a commission as a colonel in the Army Judge Advocate General's Corps. "If I wasn't patriotic enough for the District Attorney's office I certainly ought not to be patriotic enough for the Army." Wheeler, page 163.

51. Wheeler, page 163.

52. *Helena Independent,* October 19, 1918, page 4.

53. Wheeler, page 164.

Chapter 25. "Down Deep"

1. Quoted in the *Montana Standard,* May 19, 1982, page 3.

2. Burton K. Wheeler, *Yankee from the West* (1962), page 177.

3. Michael P. Malone, Richard B. Roeder, and William L. Lang, *Montana: A History of Two Centuries* (1976), pages 287–288.

4. Malone, Roeder, and Lang, pages 280–283.

5. *Fourth Biennial Report of the Department of Labor and Industry* (1919–1920), page 11.

6. Incredibly, among those supporting Wheeler in 1922 was Will Campbell, editor of the *Helena Independent*. Wheeler, rationalized Campbell, had "forsaken the evil political company" he had previously kept. Charles Sackett Johnson, *An Editor and a War: Will Campbell and the Helena Independent, 1914–1921* (1977), master's thesis on file at the University of Montana's Mansfield Library, Missoula, page 254. Campbell would die in 1938. Johnson had the following concise and insightful summary of Campbell's career: "In some ways, Campbell aptly symbolized the wartime years. He was a talented man overtaken, like the nation, by an obsessive hysteria." Johnson, page 258.

7. See *Anaconda Standard,* November 8, 1922.

8. Allegations against the attorney general included that he had dropped antitrust charges against companies that had then given political donations to the National Republican Committee. Paul Hopper interview with Burton K. Wheeler, May 1, 1968, page 23 (Columbia University Oral History interview, on file at the University of Montana's Mansfield Library); Wheeler, pages 213–229.

9. Felix Frankfurter (then a law professor and later a Supreme Court justice) wrote in *The Nation,* May 21, 1924, that "never in the history of this country have congressional investigators had to contend with such powerful odds . . . and never have such investigations resulted in compelling correction through the dismissal of derelict officials." Quoted by Wheeler, page 234.

10. The Democrats' candidate, John W. Davis, would later condemn the KKK. See Wheeler, pages 251–252, 262. For election statistics and other background on the 1924 election, see AmericanPresident.org and PresidentElect.org.

11. Wheeler, page 294.

12. Ibid., pages 297–298, 302.

13. Paul Hopper interview with Burton K. Wheeler, March 18, 1968, page 25 (Columbia University Oral History Interview, on file at the University of Montana's Mansfield Library); Wheeler, pages 304–305.

14. See, e.g., Ted Morgan, *FDR* (1985), pages 421–423, 431–432.

15. Malone, Roeder, and Lang, pages 296–301.

16. The most complete account of this trial, including extensive quotations from the transcript, is found in George Harrison's *The I.W.W. Trial, Story of the Greatest Trial in Labor's History by One of the Defendants* (1969, original edition, 1919).

17. *Fourth Biennial Report of the Department of Labor and Industry* (1919–1920), page 11; Jerry W. Calvert, *The Gibraltar* (1988), pages 115–118.

18. Calvert, pages 121–122.

19. Ibid., pages 122–123.

20. Section 7(a), NIRA. The Supreme Court invalidated these labor protections in May 1935, but Congress created even more expansive labor rights in the National Labor Relations Act (or Wagner Act), signed by President Roosevelt on July 5, 1935. See, e.g., Morgan, pages 420–421.

21. See "The First Sixty Years" on nlrb.gov. See also Milton Derber and Edwin Young, *Labor and the New Deal* (1957), pages 8–10, 204–208.

22. Malone, Roeder, and Lang, pages 328–329.

23. *Butte Daily Post,* May 8, 1934, pages 1, 8. The *Post* also attacked the bona fides of the union, claiming that the vote to strike had been taken without the presence of most members. It attacked the integrity of the strikers, claiming that the miners lacked the decency to leave some workers on duty to attend flood control and fire prevention. "Never before in the history of labor difficulties in Butte has such an attitude been resorted to by labor organizations."

24. See *Montana Standard,* September 20, 1934, pages 1–2.

25. Ibid., page 2; Malone, Roeder, and Lang, page 329.

26. Wheeler, page 319.

27. Ibid.

28. See, e.g., Frank Freidel, *Franklin D. Roosevelt, A Rendezvous with Destiny* (1990), pages 231–238; Morgan, page 472.

29. Wheeler conversation with Hillman, cited in K. Ross Toole's *20th Century Montana: A State of Extremes* (1972), pages 192–193.

30. Wheeler, page 325.

31. Alpheus Thomas Mason, *Brandeis: A Free Man's Life* (1956), pages 625–626.

32. Ibid., page 626. Wheeler tells the same story in his autobiography, pages 327–328.

33. Wheeler, pages 327–329.

34. Mason, page 626; Wheeler, page 332.

35. Morgan, page 473, referring to then-Assistant Attorney General Bob Jackson.

36. Wheeler, pages 334, 339.

37. Ibid., page 19.

38. Paul Hopper interview with Burton K. Wheeler, March 22, 1968, page 27 (Columbia University Oral History Interview, on file at the University of Montana's Mansfield Library); Wheeler, page 19. FDR had certainly been sensitive to this widespread isolationist tendency, making repeated pledges of nonintervention to the American people. As late as 1940, FDR would say, "I have said this before, but I shall say it again—and again—and again—your boys are not going to be sent into any foreign war." Wheeler did not believe—

as do some revisionists—that Roosevelt deliberately manipulated events to draw the country into war. Rather, he feared that the president's policies put the United States on a path that would lead inevitably to its entanglement in the war—whatever his intentions to stay out.

39. Wheeler, page 27.

40. In the spring of 1940, rumors began to circulate that FDR would ask Wheeler to run with him. On multiple occasions in the months leading up to the convention, the White House apparently used a variety of backchannel emissaries to determine if Wheeler would serve. Paul Hopper Interview with Burton K. Wheeler, March 22, 1968, pages 5–7 (Columbia University Oral History Interview, on file at the University of Montana's Mansfield Library); Wheeler, pages 362–368.

41. Wheeler, pages 367–368.

42. Wheeler, interestingly, had a close relationship with Harry S Truman, acting as his mentor when the freshman from Missouri was assigned to Wheeler's Interstate Commerce Committee. See David McCullough, *Truman* (1992), pages 213, 254.

43. Malone, Roeder, and Lang, pages 312–313.

44. Paul F. Healy, Introduction, *Yankee from the West* (1962), page ix.

45. Wheeler, page 430.

46. *Wall Street Journal,* July 27, 1923, page 3.

47. Though they would never again approach their World War I level of production, the North Butte properties were still mined sporadically between 1923 and 1950, when they were purchased by Anaconda.

48. For a general (and highly uncritical) discussion of Anaconda in the 1930s and 1940s, see Isaac F. Marcosson, *Anaconda* (1957), Chapter 10: "Through War and Depression."

49. Marcosson, Chapter 9: "Anaconda in Chile." The profit figure is from Janet L. Finn's *Tracing the Veins: Of Copper, Culture, and Community from Butte to Chuquicamata* (1998), page 65.

50. See Marcosson, page 273.

51. Marcosson, page 206.

52. The best and most informative writing about the Berkeley Pit is Edwin Dobb's remarkable article, "Pennies from Hell," *Harper's Magazine,* October 1996, page 39.

53. On Allende, see Finn, pages 9–10. On the demise of Anaconda, see Malone, Roeder, and Lang, pages 325–327.

54. Malone, Roeder, and Lang, page 327.

55. *Montana Standard,* September 10, 1980, pages 1, 8; *Montana Standard,* September 30, 1980, page 1.

56.   Author interview with Don Peoples, November 10, 2004.

57.   *Missoulian,* March 27, 2002, page A2.

58.   *Montana Standard,* May 19, 1982, page 3.

59.   "Rebirth of a Mining Town, Maybe," *U.S. News & World Report,* October 6, 1986, page 24. See also Jim Robbins, "Mining Again in a Montana Town That's Fallen on Hard Times," *New York Times,* November 8, 2003.

60.   Malone, Roeder, and Lang, page 323.

61.   Bill Richards, "Power Outage: For a Montana Utility, A Gamble on Telecom Looks Like a Bad Call," *Wall Street Journal,* August 22, 2001, page A1.

62.   See Kris Hudson, "Montana's Power Failure—High Tech Meltdown Meant Doom for Touch America," *Denver Post,* August 10, 2003; Richards at page A1.

63.   Author interview with Jim Keane, September 23, 2004.

Epilogue: "Normal for Its Time"

1.   *Glasser File,* pages 77–78. The *Glasser File* was a report prepared between 1933 and 1934 by Ira Glasser, a civil servant at the U.S. Department of Justice. Glasser's report, which was not published, was commissioned by President Franklin Roosevelt. Roosevelt feared that the severe economic conditions of the Depression might lead to domestic unrest, and wanted a study of the federal government's past handling of such situations. One focus of Glasser's study was Butte during and after World War I. A copy of the *Glasser File* is available at the Butte-Silver Bow Public Archives.

2.   Information and quotations from Manus Dugan Banko are from three author interviews: March 11, 2004; June 10, 2004; and November 9, 2004.

INDEX

Speculator mine, 2
air circulation in, 11
cages in, 28, 30, 127–28, 129, 132
Duggan's group in, 80–83, 84, 87,
98–104, 116–21, 148, 149, 150,
154–55, 156, 161, 183
Duggan's group rescued in, 125–33,
145, 146, 157, 158–59, 167, 180,
182, 186
end of operations at, 265
Granite Mountain's connection with,
14–15, 27–28
Moore's group in, 145–52, 153–57,
159–60
Moore's group rescued in, 161–68,
169
morgue established at, 110, 111–12,
158, 163, 183
1914 violence and, 141–42
shaft repair in, 30
shafts and tunnels of, xii, 14–15
2,000-foot level of, 32
2,200-foot level of, 31, 145–47, 164,
166
2,400-foot level of, 31, 75, 77,
80–83, 123–27, 130, 132–33,
146, 183
2,600-foot level of, 31, 100, 168
see also North Butte mining disaster
Spihr, Mike, 129
Spriggs, A. E., 46, 172
Standard Oil Company, 61
Amalgamated formed by, 61–62
Anaconda stock and, 61–62
competition and, 85, 138
Daly's sale of Anaconda to, 60–61,
65, 66
. Heinze vs., 60, 66–72, 73, 138, 139,
197

mining properties consolidated by,
72–73, 84–85, 136, 137
minority shareholder suits against,
70–72
see also Anaconda Copper Mining
Company
Stegner, Wallace, 254
Steinberg, Herbert, 95
Stewart, Sam V., 195, 196, 219, 220,
224, 235
Stewart mine, 142
Still, Leonard, 99, 102
Stone, Geoffrey, 250
Stone-Manning, Tracy, 269
Strike Bulletin, 190, 199, 200, 202,
205, 208, 210, 211, 214–15, 238
Sullau, Ernest "Sully," 56, 59
background of, 8–9
death of, 26, 37
electrical cable and, 7, 9–10
estate of, 230
fire and, 10–11, 12, 14–15, 26–29,
31–33, 37–38, 77, 116
Sullau, Lena Benson, 8–9, 26, 38, 230
Sullivan, Mike, 166
Supreme Court, U.S., 256, 265
Roosevelt's plan for, 260–62,
263–64
Swindlehurst, W. J., 169, 196

Tammany, 40–41, 94
Tevis, Lloyd, 24
Thomas, John "Shorty," 50–51
Thompson, Mrs. B., 108–9, 228
Thompson, Vernon, 109, 228
timber industry, 42, 60, 176, 268
Toole, Joseph K., 72
Toomey, "Con," 48
Touch America, 271, 274

© Kim Tilley

MICHAEL PUNKE grew up in the West and lives with his family in Montana. He is a former partner in a Washington, D.C., law firm, and his professional experience includes work on Capitol Hill and the White House National Security Council. He is the author of a novel, *The Revenant,* based on the true adventures of a nineteenth-century frontiersman. *The Revenant* was a Spur Award finalist.